THE OFFICIAL
ROBOSAPIEN
HACKER'S GUIDE

THE OFFICIAL
ROBOSAPIEN
HACKER'S GUIDE

DAVE PROCHNOW

McGraw-Hill

NEW YORK I CHICAGO I SAN FRANCISCO I LISBON
LONDON I MADRID I MEXICO CITY I MILAN I NEW DELHI
SAN JUAN I SEOUL I SINGAPORE I SYDNEY I TORONTO

The **McGraw·Hill** Companies

1 2 3 4 5 6 7 8 9 0 DOC/DOC 0 1 0 9 8 7 6 5

ISBN 0-07-146309-7

The sponsoring editor for this book was Judy Bass, the editing supervisor was Stephen M. Smith, and the production supervisor was Pamela A. Pelton. It was set in ITC Officina Serif by Cindy LaBreacht. The art director for the cover was Anthony Landi.

Printed and bound by RR Donnelley.

McGraw-Hill books are available at special quantity discounts to use as premiums and sales promotions, or for use in corporate training programs. For more information, please write to the Director of Special Sales, McGraw-Hill Professional, Two Penn Plaza, New York, NY 10121-2298. Or contact your local bookstore.

 This book is printed on recycled, acid-free paper containing a minimum of 50% recycled, de-inked fiber.

CONTENTS

FOREWORD

June 7, 2003, 1:38 AM, Hong Kong:

I'm sitting in a cramped receptionist's supply closet. It's late, and I type bombarded by staccato neon through a third-story Hong Kong window. The rolling colors illuminate my Mac, small desk, papers, cans, food wrappers, littered floor, and my workhorse: the first Robosapien prototype.

The lights off to quiet the annoying fluorescent buzz, I'm writing detailed program specifications for my first mass-market humanoid robot, linking together the nature and structure of its brain. I move the robot into various positions like a puppet, associate these gestures to sound and actions, organize the script for sensors, motors, and balance, power-on test, then type the sequence down line by line.

My years of indulgent mass media megalomania training told me creating robots was supposed to be different. Solitary castles, lurching minions, high ceilings, maniacal laughter, maybe lightning. Not tonight. The reality is an early deadline for the programmers who'll be trying to push seventy pages of personality pseudo-code into a processor smaller than a pocket calculator.

Aside from the operating system, I'm deliberately sneaking in various routines making the Robosapien's brain useful for more than just personality. He has to be more than the sum of his parts: a bot on a mission.

I reflect on how this is a bit different from typical academic methods. My style of robotics is what is called bottom-up, which is where you build things from simple components and see if you can push them to do something exotic. This is different from the usual top-down approach which is to force a fast laptop computer to roll around in the hopes it doesn't get a disk crash. I never could get into top-down methods because they required huge costs and time. Few researchers ever cozied up to bottom-up because they claimed the behaviors were too simple.

A brawn-versus-brain issue, my Robosapien is just the latest in a long series of "evolved" designs that have shown some very complex behaviors indeed. Starting out as a simple two-transistor roller in 1988, 15 years later Robosapien 1.0 (the precursor to my workhorse friend here) has less than 30 transistors controlling eight motors in a humanoid form. Doesn't fall, moves fast, looks around, picks up, lasts for hours on batteries, and without a single line of code. It's the latest of hundreds I've built.

Brain versus brawn. A misnomer: you need both, and not just in your robot.

When I was a kid, my father brought my sister and I a box of teak deck cuttings from a local shipyard. They were matchbox sized, smooth, unpainted, perfect, with clean edges. Stacked beautiful, smelled great, tasted wonderful. I remember them because they were so quick and easy to build with—walls, houses, igloos, and bridges went together in seconds, and collapsed just as quick. A far cry from commercial block kits which needed a lot of pushing, color matching, and disassembly.

Most of my life I've been searching for a similar set of blocks for robot design—something a lot easier (and cheaper) than the digital convention. Since I was small, I've always been certain that the problem of building humanoid robots had a simple, elegant solution. Only 40 years later, quick as a flash, I've now got the opportunity to apply it. And as long as it works, my bosses at WowWee Ltd. don't care if it runs on magic pixies.

At this moment, the market for the Robosapien is far from clear, but I'm hoping some universities and colleges might notice you could teach entire courses in reaction-based robotics with it. The arms are multiposition to move things about on a step, the head is low so a camera glued to the top would be where it's eyes should go, all the internals are sealed and modular, the chest is mostly empty space for mounting additional electronics, a single Phillips screwdriver completely takes the robot apart (Fig. F-1). Each of these with the goal of providing a convenience for the tinkering few, while still being fun for the user many.

The robot will be fun by itself, but for years I've promoted maverick education robotics to see if I could find the capable few. Finding they are few, I

FIGURE F-1
Mark W. Tilden and the best-selling robot of all time, Robosapien. (Photograph © WowWee Ltd.)

decided to promote skills through teaching, mentoring, and the diffuse but still ongoing BEAM robot games. Building robots is easy, but building roboticists is hard.

If I can't find them, I'll make them instead, but I'll need bait. Something shiny, with built-in skills and a less-than-serious personality. Something real that matches fiction instead of taunting it. Something that doesn't require a king's ransom to own.

I look at my prototype's glowing red eyes as it scans about in the dark. First of a species or another one for the museums?

Crazy talk. Some things are worth doing for themselves. Not for profit, not for fame, but because they're cool.

I sigh, and get back to stuffing my robot's personality into printed piles, hoping somewhere out there, there'll be hackers who get it, will have fun, and the species will progress.

Over the horizon, a roll of thunder. Nothing serious, but nice to know it's there.

Mark W. Tilden
Robotics Physicist
March 21, 2005

ACKNOWLEDGMENTS

It has been said that a book is a collage of contributions housed inside a cover of singular authority. While this book has definitely benefited from many varied contributions and collaborations, the ultimate success (I actually consider this book to be a resounding shared success, but I'm kinda biased in my opinion) or failure of this work rests solely on my neck.

With that noble gesture aside, there is a significant list of folks who worked tirelessly to ensure that this book was the best assembly of Robosapien hacks. Amelia, Amy, Anthony, Katherine, Kathy, Judy, Mark, and Penelope each generously contributed their time, hard work, and knowledge to make *The Official Robosapien Hacker's Guide* the best book on the subject, bar none.

Additionally, many manufacturers provided product samples for reference throughout the pages. Without such generous support, it would've been impossible to write this book. Acroname, A7 Engineering, Bluetooth Organization, BRIO Toys, Freescale Semiconductor, Gleason Research, iBotz, JoinMax Digital Tech, Magnevation, MEMSIC, MGA Entertainment, NetMedia, New Micros, OWI Robot, Parallax, Robotics Connection, Winbond, and WowWee Ltd. each helped immeasurably with the ultimate goal of hacking Robosapien.

Thank you one and all.

Dave Prochnow

ABOUT THE AUTHOR

Dave Prochnow is a frequent contributor to *MacAddict*, *MAKE*, *Nuts and Volts*, and *SERVO Magazine*, as well as the award-winning author of 25 nonfiction books including the best-selling *Experiments with EPROMs*. He also won the 2001 Maggie Award for the best how-to article in a consumer magazine. In collaboration with Mark Tilden, Dave has assembled the enormous selection of robot tips, programs, and hacks that are contained in *The Official Robosapien Hacker's Guide*. You can learn more about this book and other robotics/electronics projects at Dave's Web site, www.pco2go.com.

Nits in the Beard of the Creator

Rarely today do you find an engineer enjoying the same cultlike groupie status that is more commonly associated with athletes, entertainers, and contemporary musicians. This omission is indeed a regrettable oversight from an overindulgent, hedonistic society. Oh, sure, Bill Gates, Dean Kamen, and Steve Wozniak are familiar names to the average quarter-pounder mentality, but popular folk heroes are *not*. And therein lies the quandary: Society adores celebrity figures or occupations that most people will never attain in real life, while abhorring a profession that could easily fulfill their lifetime hamburger needs. Or, as stated by Dean Kamen on his DEKA Research and Development Corporation Web site (www.dekaresearch.com): "You have teenagers thinking they're going to make millions as NBA stars when that's not realistic for even 1 percent of them. Becoming a scientist or engineer is."

Yes, Mr. Segway has a mission. He wants you to become a scientist or an engineer. Although a noble goal, it is, unfortunately, a pipe dream. College is a tough, if not impossible, aspiration for most people—heck, Kamen, himself, didn't even graduate. More realistically, however, Kamen offered a better grounded and accessible approach to achieving this goal in a September 2000 interview in *WIRED* magazine (Issue 8.09; www.wired.com/wired/archive/8.09/kamen_pr.html). Kamen sees America's secondary school science education system as a dysfunctional effort that typically results in school districts throwing big money into computers, teachers' salaries, textbooks, and Web access with limited, if not unreliable, results. Another, more practical approach proffered by Kamen is to have students build robots; more specifically, "they need to have access to challenging, hands-on projects that result in a tangible product." And as a finishing touch, Kamen wants local engineers and scientists to serve as mentors for U.S. schools so that students associate a face with science.

Remarkably, this approach seems to work. Kamen's For Inspiration and Recognition of Science and Technology (FIRST) teams corporate America with budding engineers. This collaborative effort culminates in a yearly robot fest (www.usfirst.org), which Kamen calls "the NCAA of smarts." In spite of these achievements, FIRST remains a classroom endeavor, and as such, it is comparable to a specialized club project that won't appeal to mainstream kids.

A more practical and pervasive solution would dwell subconsciously in the popular world of entertainment, video games, and music. This approach would excite kids into wanting to learn about science and technology through involvement with their games, music, and toys. Toys? Yes, toys? Imagine a toy filled with electric motors, a voice, and personality. Imagine a toy that you can program from the comfort of your living room sofa. Imagine a toy that can walk, has movable hands, and can pick up your socks. Imagine Robosapien.

Officially announced at the February 2004 International Toy Fair in New York City, Robosapien became a best-selling toy during the subsequent Christmas 2004 shopping season. In fact, some sales figures reported sales exceeding 1.5 million units sold during the holiday season, alone. Not bad for a science/technology toy with a retail price of $99.95. The real story behind Robosapien, however, isn't the black and white glossy plastic humanoid robot. Rather, it's Robosapien's virtual father—Mark W. Tilden.

Although most engineering talents like Gates, Kamen, and Wozniak are nerdy types who hardly inspire adulation among youngsters, Tilden (Fig. I-1), though, is a man's man—cue Monty Python's Flying Circus' "Lumberjack Song." He's big, affable, creative, and opinionated, just the type of person you'd like to be your science mentor.

Anyone with even a passing interest in robotics knows about Mark Tilden and his robots. Thanks to his revolutionary research at Los Alamos National Laboratory in New Mexico, biomorph, Unibug, and BEAM (biology, electronics, aesthetics, mechanics) have become elemental building blocks in every budding robot builder's vocabulary. Likewise, Tilden's zeal for all things analog have endeared his twitching critters to all AVR, PIC, and RCX builders who have ever attempted to program natural movements into their own digital creations.

The landmark book *Robo sapiens: Evolution of a New Species* by Menzel and D'Aluisio is scattered with many of Tilden's offspring and carries the underlying theme for much of his research work: Could robots and humans meld into a single species—*Robo sapiens*? Oddly enough, this terrific book incorrectly overlooks two of Tilden's pet projects from the Glossary: biomorph and BEAM. Nonetheless, this valuable field guide to robotic fauna presents a revealing glimpse into Robosapien's evolution.

And this is an evolution that would make even Darwin blush.

FIGURE I-1
A rare moment for Mark Tilden—standing still. The hyper-busy BEAM builder who invented Robosapien. (Photograph © WowWee Ltd.)

Tucked away on the final page of *Robo sapiens* co-author Faith D'Aluisio's interview of Tilden are the disembodied parts for an upcoming creation—Nito (Neural Implementation of a Torso Organism). When pressed by the interviewer, Tilden declares that Nito will "interact in a simian-like fashion in its world." Even better, a photograph showing Tilden "tweaking a controller to fine-tune its response characteristics" depicts Nito's head as a red-eyed platform with a rotating neck pedestal.

Tweaking? Isn't *tweaking* the work of a creator? Is Nito really Robosapien b1? Well, let's ask Tilden and find out.

Five Questions

DAVE PROCHNOW: Honda Asimo and Sony QRIO—tour de force or tour de farce?

MARK W. TILDEN: Deep question. The answer is, of course, splunge. Why make something simple and elegant when it can be made complex and wonderful? The problem is option B requires manpower, time, and money. Most builders with the divine Frankenstein complex don't have that.

I've always known that performance to cost ratios have to be maximized if anyone is to get anywhere with personal projects. I've spent a lot of time trying to develop my own skills so they could finally be profitable. On the way, I found that when it comes to robots there are two schools of thought, top-down and bottom-up.

Top-down robot development is being spearheaded by big names like Sony, Honda, Toyota, et cetera, and all the more power to them. Their stuff is wonderful, but is unlikely to get out of their media circus.

Biomorphic robots aren't much better though, but at least they keep the costs down to the point where they can fit market expectations. There'll come a time when the benefits of both methods will overlap, then we'll start to see some real robot evolution. In the meantime, I'll keep building with humble components because it allows me to build a robot in days instead of years.

On the other hand, would I like to have the sort of finances and resource that would allow me to build oversized robot armies like a hyperactive James Bond villain? Tyrell was a slacker.

DP: Follow-up question: In your opinion, are robot designers too hung up on anthropomorphism? And, if so, does this desire to make their creations more "human-like" stymie advancement in robotics?

MWT: I pushed anti-anthropomorphism for years, and though I made many robots, I didn't make many fans or customers. The natural shape of a robot descends from its function, so I built utilitarian robots by the score. Insects, for example, are the ultimate example of a robot Swiss-Army knife. They're capable in many environments, but it's hard to feel empathy towards them. People respond well only to things they recognize, and the mass media has made people prefer robots with human-like attributes.

Pity, as I've always been interested in the bizarre and unusual designs possible. Biologics only come in a small variety of round, tubular, flat, and bilaterally symmetric arrangements, whereas robotics allows us to build any design outside of convention or scale restrictions. We could build the fantastic, and I hope someday we will. But for now, we have to eat, so you'll see many human-form machines in the coming years.

DP: Follow-up question: I have three daughters, and they all reacted similarly upon first viewing Robosapien—fear and trepidation. And these children aren't robot newbies. Sure I had made several wheeled, tracked, and rolling robots over their collective life spans, but none of my contraptions had approached the lifelike realism of Robosapien. Shortly after the dissipation of their fears, each child adopted a unique reaction to Robosapien. For example, the 5-year-old treated it like a companion, the 3-year-old considered it to be her "Roro" playmate, and the 1-year-old mimicked its walking and body movements (as an aside, the 1-year-old stole the Robosapien remote control and programmed a complete 14-step master program—through happenstance, no doubt—while I was reading the user's manual). What is it about Robosapien that causes this visceral reaction among children that is absent among adults?

MWT: Grownups see the robot as a small but harmless toy, children see a fast-moving indeterminate hulk almost half their size. It's some sort of caution leftover from our caveman DNA, I think. Still, who can blame them.

Years ago, I saw the response of people and animals to various autonomous robots I built. It led me to invent the *Purring Test* convention. Many will recognize the *Turing Test*, which measures wether a human can tell if a computer is intelligent. The Purring Test is what a robot has to do to make a cat think that the robot is alive.

This came out of many years directly observing (and repairing) such relations. My cats seemed to have an innate understanding if something is being controlled by the annoying human, or if it is truly self-guided. When I set out my first solar-powered robot floor cleaners back in the late 80s, the time-lapse video was very interesting. Robot moving in short jumps every 6 minutes, large black obscurity with claws, then hours of a robot's carcass with wires torn across the floor. What I learned (aside from the fact that my cats had a 20-minute pounce patience) is that if robots are going to live in real environments, they have to be designed with thought to the current ecological occupants.

The Purring Test is still a standard I use today. It's a general measure of interactive competence with the real world outside of any form of comprehensive intelligence. For example, Stephen Hawking, regarded as one of the most intelligent people on the planet, couldn't pass it. Makes ya think.

But what makes the Robosapien seem so alive is that he's not based on precise feedback motions or orthogonal limb actions. Almost every other robot looks like it's been designed on graph paper; my robots are based more on biological triangles. This gives them a range of motion that, though efficient, is a bit more interesting to look at for some reason.

For myself, I'm just trying to optimize performance given limited motor dynamics, but if other people see this *aliveness*, then it's a bonus I like to think typifies the biomorphic design style. The real advantage is that kids see it immediately. "Of course it walks, it's a robot" was one of the best compliments ever given the Robosapien from a small boy. He didn't question it; he just accepted it, even though casual walking is one of the hardest things to do in robotics.

Purr.

DP: Follow-up question: Then, is Robosapien the culmination of your 212-transistor Nervous Integration of a Torso Organism [sic] (Nito 1.0) compliant anthropoid described in your seminal 1995 paper, "Theoretical Foundations for Nervous Nets and the Design of Living Machines"?

MWT: The original Nito was to be a lot more sophisticated. The original Robosapien (still working today) was completely run from a 28-transistor 4-tricore Nervous-Net array, quickly patched together to meet deadlines. Nito and RS1 are architecturally and ambitiously different at many levels.

Over several years, my colleague Brosl Hasslacher and I had developed a series of self-optimizing Nervous Network architectures with significant performance-to-silicon ratios. We soon realized that these could be structurally fashioned as we like, and Nito was to be our anthropomorphic opus. Alas, government funding sources didn't see it, so I started it using my own time and resource, but soon it had to take the back burner to paying projects.

It still sits there in the future, though: a self-optimizing, gold-plated, adaptive humanoid in less than 212 transistors. An optimization of form, function, and physics.

Maybe someday I'll complete it for the same reason they finished the Babbage engine. Obsolete, but what a beauty.

DP: In a November 28, 2004, interview in *The New York Times*, you were quoted as saying that "robots are the last unfulfilled promise of the 20th century." What about the jet pack, hovercraft automobiles, and meals in a pill? I'm still waiting for these, aren't you?

MWT: Yeah! And what about wall-sized TVs, and pocket phones, and a hundred TV channels, and a worldwide graphic information network, and....

Hold on.

Yes, the 60s dreams of the future are still in demand, but I've always realized that we'd have to build them, not wait for them. The problem with all the other dream items is that they require funding, help, finances, politics, and risk currently beyond the enthusiast. Robots aren't like that. Robots are the only future technology where everybody knows what it is; we just don't know how to build it yet.

Welcome to the year 2000-and-change. We are the future, so we'd better get good at it.

DP: Do you know Hans Moravec (cofounder of Seegrid Corporation)? He has predicted that enhanced supercomputing muscle could result in "self-aware, emotional" robots by 2040. What do you think an emotional, self-aware robot would think about its progenitor being sold as toys?

MWT: It would probably resent it as much as humans resent being prehistoric saber-tooth jerky (i.e., not at all). We got the last laugh, though. I doubt if robots will someday do the same.

We don't have to worry that emotional robots will someday replace us. Don't forget that humans share the planet with about 20 million other species, all with the same evolutionary breaks and motivations we've had, yet not one of them has made a play for dominance.

Why would machines? What's their motivation? Hunger? Sex? Money? Power? You need to be based on cellular protoplasm to make these vital lies important enough to get out of our leather recliners.

You will never have to worry that your toaster will one day force you to sleep in the guest bedroom. Their thoughts will be glacial, not vindictive, by design.

DP: Follow-up question: Could Hasbro's ill-fated B.I.O. Bugs [B.I.O. stands for Bio-mechanical Integrated Organisms] marketing fiasco in the post-9/11 toy industry climate been avoided by incorporating a "hacking" attribute into the concept?

MWT: B.I.O. Bugs were designed hackable, just never promoted as such. The problem is that bugs, even when significantly stylish, are just not personable enough to encourage the hacking gene. As for the marketing fiasco, any sales following 9/11 were...difficult, despite marketing attempts.

B.I.O. Bugs were a great market lesson for me, though. The average complaint was "I hate the noisy thing; my kid has taken it everywhere for two months." As a designer, I was glad my target audience liked the product, but as a business head, I'd obviously missed the primary repeat buying target, i.e., the parents.

One idea was to make new robots so they behaved quietly, like portable game-stations. My approach was make our new robot so it would be attractive to grown-ups, too.

Seemed to work.

Company Profile

WowWee Ltd. was founded in 1988 as an independent research and development and manufacturing company, focused on incorporating affordable, cutting-edge technologies into toys and other leisure time products to provide enhanced entertainment value.

In the early years, WowWee was an OEM seller, to retailers and to distributors. Among the products developed and manufactured by WowWee and marketed under other brand names at that time were Power Ranger Power Gloves and Talking Tots dolls (Happiness Express) and Animal Planet Animaltronics (Toys "R" Us private label).

In the late 1990s, WowWee established a sales organization and began to market products under the WowWee brand name, including the Dinotronics RC T-Rex, Totally Extreme Skateboarder, and Megabyte Dog.

Today, WowWee, while pushing the limits of technology, focuses on creating breakthrough consumer electronic and leisure products that set the standard for innovation, with unique qualities that distinguish them from the pack. The company is in the process of expanding its line of stylish and sleek Robonetics entertainment robots, as well as developing a revolutionary speech driven interface for the computer, as it creates a niche in the marketplace for affordable high-end gadgets.

WowWee is a privately owned, Hong Kong–based company, with offices in North America and a worldwide sales and distribution network. Company divisions are WowWee Robotics, WowWee Tech, WowWee RC, and WowWee Alive.

DP: In an interview from a March 2000 exhibit of your robotic design art (you do consider yourself to be a "transistor artist," don't you?) at the Bradbury Science Museum in Los Alamos, New Mexico, you mentioned that you envisioned a "future" house that would be inhabited by "home-cleaning robots." In fact, you believed that we (us human beings) would buy and sell our houses without taking these robots with us; or, as you likened, leave them behind like the light bulbs. Does this concept of a house infested with robots blur the distinction between the robot and the cockroach?

MWT: The cockroach is a pest, the robot ecosystem would provide a self-sustaining benefit for the homeowner. I know that because I lived it for 10 years.

At one time I had over 60 autonomous solar powered things roving my house and garden. Flexible, self-charging, slow, elegant, dishwasher safe. I can't tell you how nice it was to come home to a bachelor apartment after a week away and find it cleaner than when I'd left it.

However, there were problems with the marketing concept. Some of the machines were fairly insectoid in appearance, others were strange to guests. When my mother visited, I had to put all of them away as she was terrified of them crawling about her as she slept. I lost a date when my snakebot remote control slithered its way out between the couch pillows. How many rolls of toilet paper does your autonomous vacuum have to eat? The stories go on and on.

Problem is that even when the robots are ready, the humans are not. You have to change to a robot's patterns, because they're not capable yet of adapting to ours. It'll take a while. Have to educate the market first.

And no, I'm not an artist, though occasionally my bots were exhibited as such. I can barely draw a straight line, but I do take pride on optimizing my robots for robustness and performance. This makes them clean and functional, and others sometimes appreciate that outside the utility.

I've always believed there's a biological basis to aesthetics, that we admire the way nature builds because its function is so linked to the form. An aesthetic basis for robotics is harder to justify, seeing as how the majority of industrial robot machines would hardly win any beauty contests.

Still, maybe one day there'll be robot custom shops like there are for motorcycles, vans, and cars now. Then we might see some robot swimsuit competitions where you can take home the winners.

That'd be cool.

DP: Follow-up question: I'm reminded of a particularly creepy paper from NASA ("Insectile and Vermiform Exploratory Robots" by Thakoor, Kennedy, and Thakoor) that describes the government's altruistic employment of biomorph robots for exploration and unexploded ordnance removal (yet, oddly enough, fails to even reference your research on this topic; particularly your paper "Biomorphic Robots as a Persistent Means for Removing Explosive Mines"). While noble in its stated purpose, I can't help but envision "them" sneaking some of these "critters" into our homes for "national security" purposes. Does this "pest" concept inspire a redefinition of your biology, electronics, aesthetics, mechanics (BEAM) philosophy to represent biomorph robots as Bug, Eavesdrop, Analyze, Monitor?

MWT: Such papers and efforts were big in the 90s, but few went anywhere than research. Scientists like Robert Full (et al.) are only now finding out what I'd advocated for years which is that the structure of a Robobug, not its brains, are what allow for a lot of real-world autonomy and processing.

The fact is good programming will never make up for bad mechanics. Mother Nature found that out 3 billion years ago. It's just a pity she never published.

And as for self-mobile bugs, they're yet to be as quiet, efficient, or inexpensive as the originals. Not to worry. They won't be stealing cable just yet.

DP: You appear to have a unique vantage point of women in science. For example, you collaborated with Susanne Still with the Institute of Neuroinformatics, in Zurich, Switzerland, on a remarkable research paper ("Controller for a Four-Legged Walking Machine," circa 1997). There is evidence that you attempted to make your Hasbro B.I.O. Bugs accessible to girls. And in a 2003 *Taiwan News* interview you made a strong case for girls liking robots. In fact, you were quoted as saying that a proposed "FemBot" design should be able to, specifically "do simple tasks like combing her owner's hair.... [U]nlike boys, girls like to interact with their toys." Is it really that simple? Do you subscribe to the "pink mentality"—paint it pink and they will play?

MWT: "Do girls play with robots?" is still a big question in the industry. However, outside of my own colleagues and students, there are many fine XX chromosome bot bashers out there: Mataric, Grainer, Breazel, etc., all of whom can and do wield a mean soldering iron.

You have to remember that through history, a technology is never really accepted until it can provide casual utility for the female of the species. Cars, phones, trains, planes, ships, lighting, radio, all started out as crude, spindly geek projects. Only after several years did they reach a level of utilitarian transparency that made some girl somewhere say "Ooh, you have an airplane?" The rest is history, with women providing new generations to appreciate their mothers wisdom in both men and machines.

Which is why robots are so counterintuitive. They are a potentially intelligent gender all to themselves, but without all the hassle. Robosapien's girlfriend, for example, was to have a very different interface more appropriate to the cooperative play girls are generally fond of.

Girl robots for girls won't just involve a paint job. A completely different interface is required. Someday I hope to test the theory.

DP: Follow-up question: In your professional experiences, is there a disparity between men and women in science that makes women less likely to achieve "Einstein greatness," as reported in an article ("Who Says a Woman Can't Be Einstein?" by Ripley) in the March 7, 2005, issue of *Time* magazine in which biological and social (environmental) differences are proposed as the cause effect for scientific aptitude?

MWT: There are less female scientists for sure, but perhaps that's because women are generally smart enough to avoid the autistic, detail-intense focus that allows men to be such jerks about things like sports, cars, or politics.

As the base and source for future generations, I like to hope women may be genetically more concerned with vital issues, leaving the tech stuff to those of us who can't hunt mammoths anymore.

Grunt.

DP: Follow-up question: Isn't "FemBot" a male noun?

MWT: Which is why the more appropriate Greek-based term "Gynoid" should be used, but isn't as widely known (thanks to the Austin Powers movies). However, I think female-gender-appearance robots have a place in the robot marketplace as they naturally fall into companionship (no, the nice version of "companionship") roles better than utilitarian "GuyBots." A small sexist generalization, but one with undeniable market grip.

Regardless, we just have to proctor our plastic friends to a level where we'd be proud to show them to our girlfriends, because someday, some women will say "Ooh," and the market will be afoot!

> **AND, FINALLY...**
> **DP:** So what's with the hat motif?
> **MWT:** Superglue accident, 1997. Still, it keeps the alien thought-control messages to a minimum, and the sun off my anten...er, bald spot.

The Official Robosapien Hacker's Guide

This book is *it*. Inside its 12 chapters, you will find 11 of the most exciting "hacks" that can be integrated into a stock, commercial, off-the-shelf Robosapien (Fig. I-2). You won't find any smoke and mirrors here. These are actual robots that have been modified for performing a variety of clearly

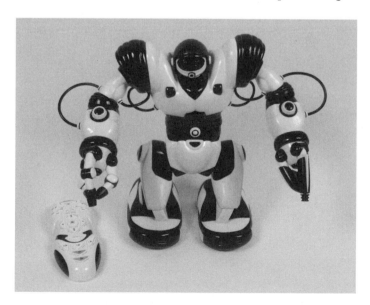

FIGURE I-2 A commercial, off-the-shelf Robosapien.

defined tasks or purposes. Every step is explained, and every detail is shown. You don't have to have solder for blood to perform all of these hacks, either. There are a wide range of hacks for an equally wide range of experience levels and hacker talents.

For example, Chapter 2 describes, in detail, how to design a new infrared (IR) remote control for Robosapien (Fig. I-3) using your computer, an IR transmitter, and some neat software. There isn't any soldering or circuitry work needed with this hack. However, if you are interested in pushing both yourself and Robosapien to the limit, Chapter 12 shows you how to add a fine new digital brain to your favorite robot (Fig. I-4). Finally, if all of these technical hacks aren't your cup of tea, then Appendix A is ready to show you how to dress Robosapien for a strut down a runway at the next Paris fashion show. If

FIGURE I-3 The Robosapien IR remote control.

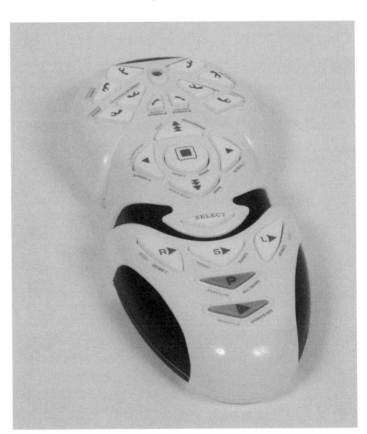

you don't want your robot to amble down the runway, then Chapter 7 will show you how to give Robosapien some serious new footwear.

Each hack how-to in this book follows a standard format. Following a brief introduction to the proposed hack, "Hack Highlights" will list the 411 for that chapter's primary project. A quick review of this list will quickly let you know if both your budget and talents are adequate for making this hack work. After this introductory information, the real work begins. A step-by-step discussion will hold your hand all of the way through the hack—from start to finish. And it had better work, right? Well, if it doesn't, a short troubleshooting section will help get your hack up and running. Finally, the "Rosebud..., One More Thing" section will give you some extra ideas or options for building on your hack. These ideas should help get your creative juices flowing and help you

FIGURE I-4 **Same Robosapien, new brain. (See Chapter 12 for learning how to give your Robosapien a new brain implant.)**

earn that online doctoral degree that you've been promising yourself. Just remember to acknowledge all of the "little robots" when you accept your award.

Above all else, in each of these hacks there is plenty of room for experimentation. Likewise, there isn't one *best* approach to a hack. Different components, different skills, and different hacker creativity will result in different hacks. So if you know a better way to achieve the same results, feel free to go your own way. Robosapien won't care. Just remember what Tilden said during a November 28, 2004, interview in *The New York Times:* "This baby is designed to be hacked." While some might attempt to disparage Robosapien by likening it to a toy, remember that this robot is 100% pure science. So what if it's a toy? In this same interview Tilden comments that he left his lucrative government employment "because there are no children to make toys for on Mars." Now that sounds like a mentor heralding science from the mountain top. And Tilden's toys go a long way toward convincing a kid that science is pretty cool.

CHAPTER 1

Crack 'n Hack

I can still remember my first robot epiphany like it was yesterday. You know what an epiphany is—that sudden realization that everything you previously had been told was just a bunch of horsefeathers, and like a bolt of lightning out of the blue, you are struck with the simple and striking *true* meaning of reality. This intuitive grasp of reality can be applied to just about everything in life where you slap your forehead and exclaim, "Aha!"

In this case, my first robot epiphany occurred on a hot summer afternoon in 1968. I was waiting for an orthodontist appointment in Omaha, Nebraska, when I begged my mother to make a momentary diversion into a local hobby store. It was there, during my hurried study of model airplane kits, that I caught sight of that essential element long missing from my experiments with robots. Like some celestial reflection proclaiming the arrival of a messiah, a rotating sign high overhead shouted, "Control Your Model, 100s of Feet Away." Sitting several feet away from this magical beacon was the device that would forever change my experiments with robots.

The Pulse-Commander was a beautiful metal transmitter that could mysteriously twitch a distant servomotor through modulations of invisible radio waves. Wow! That was exactly what I was looking for to cut the restrictive power cord umbilical that had, up until that point, handcuffed my robot to the local power outlet. My first robot hack.

At that time, my robot consisted of a 2-foot-tall "bucket of bolts" that I had fashioned from a surprising combination of Erector® set parts and some leftover plastic model airplane and armor kits. On the four-wheel, tandem-axle base of this robot, I had attached a powerful electric motor and gear set. The direction (forward or reverse) of the robot was controlled by a crude touch sensor that was attached to the motor's direction switch. Therefore, as the robot rolled along and ran the touch sensor into a wall, for example, the motor would switch direction and back the robot away from the wall. A similar directional control touch sensor was attached to the robot's rear end. Therefore, the robot would ricochet between obstacles devoid of my direct physical control. Unfortunately, there was always that irritating power cord that limited the radius of my robot's movements. The Pulse-Commander would change all of that, or so I thought.

That day of reckoning never came. As a youngster, I never had enough financing to afford such an extravagant device. Suddenly, the other realities of life intervened, and my radio-controlled robot plan was relegated to a series of notes, plans, and schematics that were similarly forgotten.

My second robot epiphany again caught me by surprise. Fast-forward to 2004. As I walked the aisles of a local toy store with nary a kid in sight, I turned to see "eye-to-LED" (light-emitting diode), a new kind of toy—a fusion of technology and personality. I saw *Robosapien*. Motors, sensors, and remote control, it was love at first sight. Remarkably, my mind immediately conjured up the query, "I wonder if I could attach a video camera to that thing?" Aha, a hack (see Chapter 4).

I didn't want history to repeat itself, so on a lark, I took Robosapien home with me. After only a cursory review of the user's manual, I was able to construct short multistep programs, control the various sensors, and test the limits of the infrared (IR) remote control. Likewise, my three children were instantly enamored with Robosapien, and unlike me, they became adept at programming multistep commands via the IR remote control. I had trouble accepting the three-layered remote control system (or maybe comprehending the complex concept of layered or shifted key presses was too tough for my advanced years, or maybe the fine print on the remote control was just outside my comfortable vision range). As such, I thought that there's got to be a better way to control this thing. Aha, another hack (see Chapter 2).

This crazy robot was a veritable cornucopia of hacking opportunities. As I set about exploring all of the various possibilities for hacking a Robosapien, I realized that this type of stuff would be extremely valuable to all of those other robot owners who were ready to take the next step—hack their Robosapien.

Once you accept this responsibility—that you will be altering a perfectly good product and, quite possibly, making a pile of mangled black and white plastic junk that will eventually find its way into the trash can—then you have become a hacker.

A hacker, me? No, you're kidding. I don't want to be arrested or something. Well, that could be kind of exciting. Is it safe? Is it expensive? Well, that does sound kind of exotic. All right, what exactly is a hacker?

First of all, forget the media's portrayal of hackers and, for that matter, the justice system's current definition of a hacker. Hackers are *not* vandals. Vandals are vandals. A crook who uses a computer or electronics to achieve an ill-gotten booty is a thief, just like the clown with a handgun who robs the local convenience store. A crook is a crook.

Hackers are visionaries who aren't satisfied with the status quo. A hacker is an artist who paints in analog and digital circuits. Hackers are environmentalists who reduce, reuse, and recycle old electronics. Always thinking, a hacker is the inventor who will make a round wheel out of a square hubcap. Our next insanely great thing will come from the mind of a hacker. Hackers *rock*.

For the hacker, a product like Robosapien is a beginning, a springboard, a foundation for building the ultimate alarm clock (see Chapter 5), a mobile sound system (see Chapter 6), or an omnipresent message board (see Chapter 10). My God, I'm pumped. I want to hack this thing! But where do I start? How do I get inside this beast and find out what makes it tick?

HACK HIGHLIGHTS

THE HACK: Learn how to disassemble a Robosapien.

WHAT YOU WILL NEED: A Phillips head No. 0 screwdriver. Optionally, Fig. 1-1 shows some other tools that will make your hacking life much easier.

HOW MUCH: $1 to $50.

TIME HACK: 2 hours.

SKILL LEVEL: Enlistee.

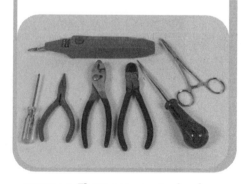

FIGURE 1-1 **These are your tools of the trade—the hacker's trade. The reamer, the second tool from the right, is vital for enlarging holes in plastic.**

The Hack

STEP 1: GET READY. OK, take a deep breath—you are about to become a hacker. Yes, lock the windows and bar the door; if the local newspaper finds out that a *hacker* is in the neighborhood, you will most certainly be arrested. Not really, but news accounts of the hacker community do paint a very ugly picture of the robot experimenter or hacker. So (un)screw 'em, grab your Phillips head No. 0 screwdriver, and let's get to work.

STEP 2: NO POWER, PLEASE. Your first step is to remove the D-size batteries from the Robosapien feet (Fig. 1-2). (*Note:* Remember to always remove the batteries before you perform *any* of the hacks described in this book. Failure to follow this simple, and obvious, rule could result in damage to Robosapien, loss of your hack's components, and a possible shock to you.)

FIGURE 1-2
Always remove the batteries before disassembling Robosapien.

STEP 3: GET COMFY. Make your patient comfortable and lay the Robosapien down on its front. In other words, so that that back, where the power switch is located is up as shown in Fig. 1-3.

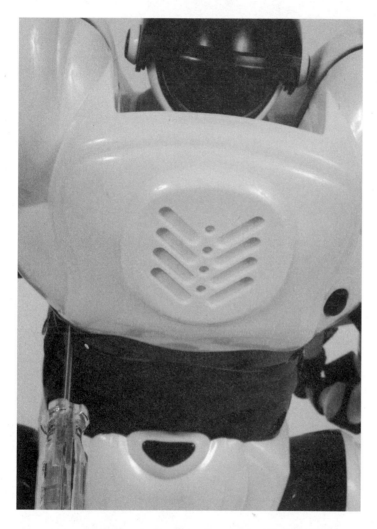

FIGURE 1-3
Lay Robosapien on a soft cloth while removing the screws.

STEP 4: REMOVE THE BACK PLATE. There are four screws that hold the back plate to the Robosapien body—one in each shoulder and two in the hip (Fig. 1-4). Once you remove these screws, the back plate will lift off. Be careful, however; the power switch wiring harness (this also holds the speaker wiring)

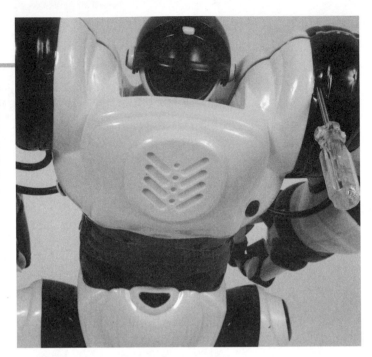

FIGURE 1-4
There are four screws that hold the front and back chest plates together.

FIGURE 1-5
Watch that wire. Carefully unplug the power switch and speaker wiring harness from the back chest plate.

as shown in Fig. 1-5 is attached to the main circuit board. Just pull the main board plug for the power switch harness and the back plate can be removed completely. Set it aside.

STEP 5: MARVEL AT THE COMPLEXITY OF SIMPLICITY. Inside the chest cavity (Fig. 1-6) are several flashes of engineering brilliance: strain relief sockets for the external wiring, counterbalancing restraint springs, a main circuit board completely populated with sockets (Figs. 1-7 and 1-8), and a cavernous chest cavity for holding all of your future hacks.

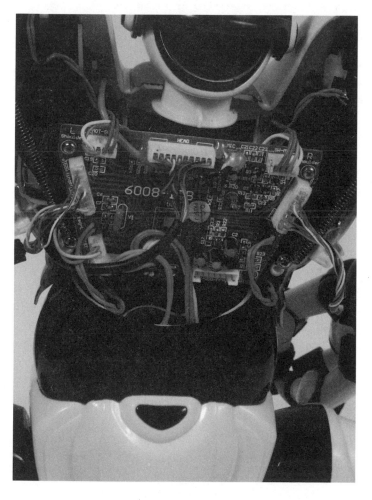

FIGURE 1-6
Hello, in there. Yes, Robosapien keeps its brain on its back. On the flip side (inside the front chest plate), there is a lot of space for holding various extra components.

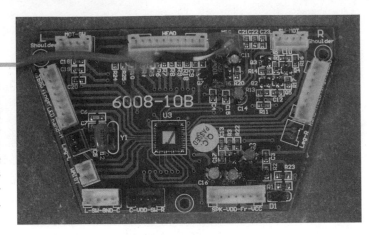

FIGURE 1-7 **The Robosapien main circuit board. The wire running along the upper side is the microphone cable.**

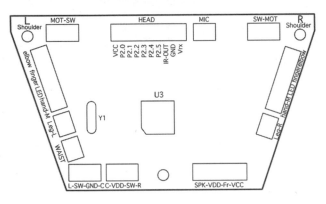

FIGURE 1-8 **A simplified plan of the main circuit board. Pay attention to the capacitor crystal, Y1. This component can be easily hacked with some great payoffs, as described in Chapter 11.**

STEP 6: DROP ITS DRAWERS. Two screws are hidden under black plastic plugs on the backside bottom plate (Fig. 1-9). Just pop the plugs and remove the screws. You might need a small knife blade for removing these plastic plugs. Be careful that you don't mar the Robosapien plastic, however. This is probably the hardest, and most dangerous, step in the disassembly of Robosapien. So take your time. In my disassembly of countless Robosapiens, I have seen easy plugs and hard plugs. Furthermore, I've seen some plugs that appear to be glued in place. (*Note:* If you have a hand drill, one of the safest and least harmful methods for removing these nasty plugs is to drill them out. Don't use a power drill for this step—too much speed will melt the plastic. Just a simple hand twist or two, and you'll have an easy access hole through each plug.) So, if you happen to scratch or mar the black plastic, take heart. This plastic is actually painted black and can be quickly recovered with some gentle sanding and a little dab of black gloss paint. Finally, there is one last screw holding the backside tight. It's, er, between the legs (Fig. 1-10).

FIGURE 1-9 **After you remove the black plastic plugs, you can get access to the two main screws which hold the backside plates together.**

FIGURE 1-10 **The final screw for the backside plates has been removed and the rear plate lifted off.**

STEP 7: SOME FACE TIME. Now turn Robosapien over, so that the chest is facing up. Remove the upper and lower front plates. The top plate should just lift off. The bottom plate, however, is held in place with two white plastic wire washers. These washers hold the two hip motor wires during assembly. They both will easily pop off with a fingernail, but make sure that you don't lose them (Fig. 1-11).

FIGURE 1-11 **Don't lose those plastic washers. They are great for keeping the motor wires in place during reassembly.**

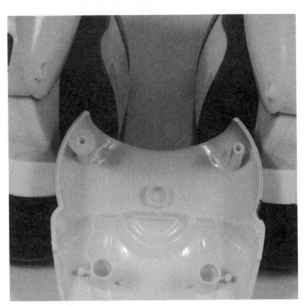

STEP 8: I'M NAKED AND FAMOUS. There are four screws for each upper thigh plate that are located on the inner thigh region, front and rear (Fig. 1-12). Two of the screws are easily found along the front and rear edge of the plate (Fig. 1-13). The two remaining screws are located very close to the movable joint and can be tough to remove. Try moving the leg forward for gaining clearance to remove each screw. I typically leave the inner thigh plates attached to the main body's hip gear boxes (Fig. 1-14).

FIGURE 1-12 **Is it just me, or is it getting cold in here? The thigh plates enclose the hip servo motors.**

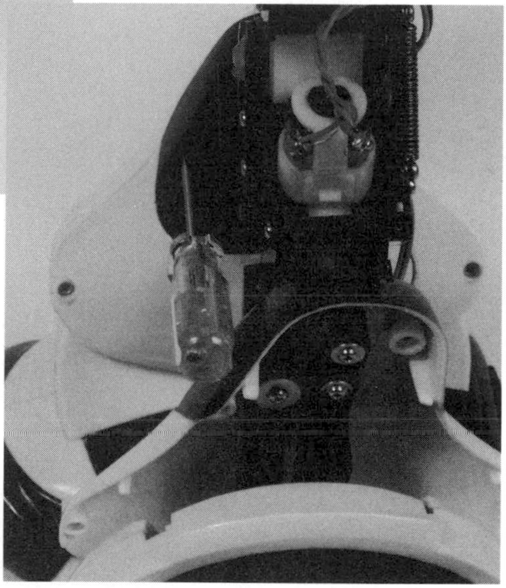

FIGURE 1-13 **Four screws hold the thigh plates together.**

FIGURE 1-14 **I leave the inner thigh plates in place during most of my hacking experiments.**

STEP 9: HAPPY FEET. This is another tricky disassembly step. There are four screws holding the feet together—two very obvious ones located on the instep and two screws that are neatly hidden by the inner upper thigh plate that you *didn't* remove in the previous step (Figs. 1-15 and 1-16). If you elected to remove these plates, then the removal of these two screws will be a snap. Otherwise, you will have to gently rock the opposite leg out of the way for gaining access to each of these screws (Fig. 1-17).

FIGURE 1-15
The battery compartment takes up most of the space in each foot.

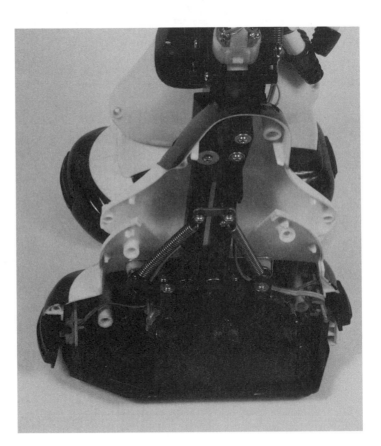

FIGURE 1-16 Wow! What happened to your toenails? Each foot touch sensor is a plastic button that engages a dual flexible rubberized sensing pad.

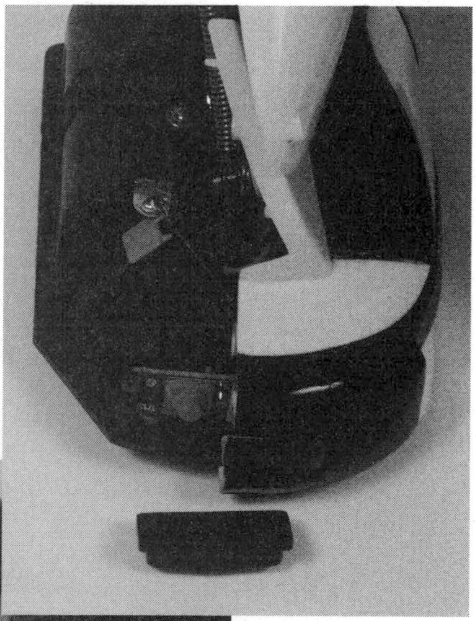

FIGURE 1-17 With all of the external plates removed, you can clearly see, and gain access to, the hip servo motors.

STEP 10: I SHOULD'VE HELD YOUR HAND. Don't remove the upper arm—only the forearm plate is removed. This limited removal will give you complete access to the hand and wrist gear box without dropping the entire arm assembly out on the floor (Figs. 1-18, 1-19, and 1-20).

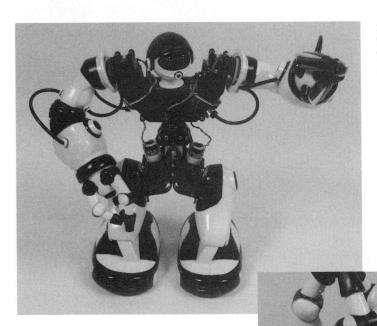

FIGURE 1-18
The arms can be difficult to disassemble.

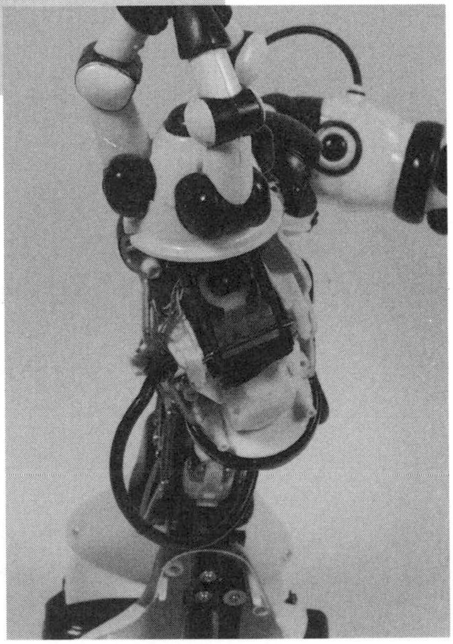

FIGURE 1-19 If the upper arm plates are left intact, the forearm plates can be removed for working on the hand, fingers, LED, and touch sensor.

FIGURE 1-20 **This beauty is more than skin deep. In fact, the real beauty starts once the skin is removed.**

STEP 11: THAT'S IT! Congratulations—you now have an expensive pile of engineering ingenuity. Try to reverse your steps and put Robosapien back together again.

STEP 12: LEAVE ONLY FOOTPRINTS. Now get ready for your own robot epiphany. Add one of the many hacks described in this book, or make your own Robosapien modification. But above all else, hack it.

Erect Your Own Robosapien

I am continually mystified why some robot builders insist on learning advanced metallurgy and metalworking skills. Oh sure, you learn how to make some great custom-designed components for your next robot chassis, but at what cost? Well, I'm a little more "old school" than that—I want to build a robot now, not waste my time bending, breaking, welding, and grinding. I look for simple parts and found items that can be easily modified or altered into my design paradigm.

One of the best sources for finding simple parts that can be quickly fabricated into a wide variety of robot chassis designs is the Erector set (the exclusive trademark of Meccano® S.N. of France). You remember Erector sets, don't you? What you might not know about today's Erector set (unless you inhabit the local toy store, as I do) is that these terrific metal building sets now come equipped with small electric motors and great plans for building up to 50 different models. These Erector Motion Systems™, as well as the more traditional Erector sets feature easily assembled pieces that can be readily adapted to any robot design requirement.

My favorite Erector set for use in robot building is the 40 Model Set (#8540) Erector Motion System (Fig. 1-21). No, it's not because there is a robot design included in this set. There is a robot, but the real reason is that there are more than 300 pieces and parts that I can quickly combine with a simple machine screw and nut joints for less than $80. Heck, I'd spend more than two or three times that price for a low-cost welding tool. More important, with my Erector sets (yes, I have more than one) I can rapidly create a prototype of a chassis idea, slap on a simple control circuit, and see if my new design is worthy of future experimentation. As a bonus, this Erector set comes with a handy plastic storage box that I can tuck away in the closet when I'm done experimenting. So think low-tech before you invest in that heavy, expensive metalworking machinery, and you will be able to build your robots (Fig. 1-22) in less time than it takes to grind a weld.

FIGURE 1-21
A robot builder's treasure trove—the Erector set.

FIGURE 1-22
Yea, it makes robots, too.

CHAPTER 2

IR Makes Your Brown Eyes Red

According to ancient home entertainment lore, Zenith Radio Corporation invented the first remote control in the 1950s. Prophetically, this device was code-named "Lazy Bones." While these early wireless remote controls were limited to basically turning the TV on and off, a quantum leap in technology was made in the 1980s with the introduction of infrared transmitters and receivers. Utilizing invisible photons of modulated infrared (IR) radiation, these improved IR remote controls were capable of controlling nearly every facet of a typical household TV, VCR, and home audio system (Fig. 2-1).

Today, the IR remote control has become the de facto user interface for operating virtually every home electronics product. In fact, you'd be hard-pressed to find much more than a power button on most TVs these days. Why the proliferation of IR remote controls over other forms of wireless control? Basically, IR communication technology is cheap, fast, relatively secure, and devoid of any kind of licensing requirement such as those found on radio-controlled devices. Likewise, the operation of an IR remote control is deceptively simple.

When you press a key on your handheld remote control, this device translates the digital key press into a series of electrical IR pulses that are beamed to your TV, DVD player, satellite receiver, or Robosapien. An IR receiver inside the TV, for example, then decodes these electrical IR pulses and converts them into the appropriate command for controlling the TV. Therefore, pressing the remote control's volume key is equivalent to walking over to the TV and physically changing the volume setting. That is, if you really could manually control your TV.

All is not bliss in the world of IR remote controls, however. There is a competing standard, IrDA (Infrared Data Association; see www.irda.org), and, even worse, different operating frequencies.

Consumer electronics utilize IR carriers operating at frequencies that are all over the place. Ranging from 30 to 56 kilohertz (kHz), the common frequency for today's audio systems, for example, is 39 kHz, whereas the IR carrier for Robosapien is 39.2 kHz. Oddly enough, the IR carrier for Bang & Olufsen isn't even on this typical frequency map. This high-end consumer electronics gear uses a 455-kHz signal.

FIGURE 2-1 Now you can anchor your home's entire IR remote control enterprise on your Mac—the ultimate remote control.

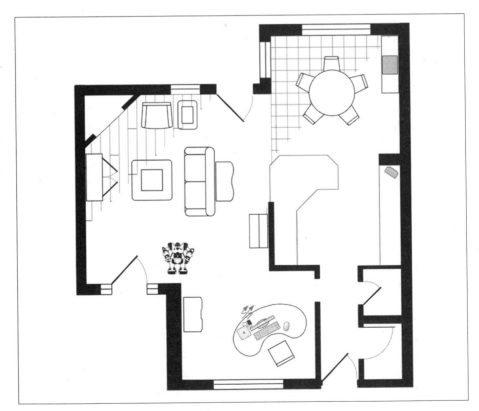

Likewise, with all of these operating frequencies, there is an equal number of different IR remote controls. As a result of this diversity in frequencies, most manufacturers bundle a unique IR remote control with every product. Therefore, consumers end up with a bushel basket full of remote controls.

Oh, and if you think that high-end manufacturers like Bang & Olufsen are immune to this type of remote control proliferation, think again. Just a simple audio system, TV, and set of powered speakers can result in your living room being cluttered with three different remote controls. That is, unless you pony up a couple of thousand dollars for a Beolink master control unit.

What you need is a learning remote control, right? Well, get ready for some sticker shock. There are two powerful learning remote controls that can master a whole household of electronics equipment: the Philips ProntoPro and the Marantz RC5200. Although fully capable of replacing all of your common IR remote controls, these stylish handheld devices don't come cheap. The ProntoPro costs $899, and the Marantz model lists for $599. Furthermore, in the case of the Marantz version, you get a squat 320 × 240 monochrome, stylus-controlled screen that is fondly reminiscent of an early PDA (personal digital assistant).

Yes, I know that there are less expensive learning remote controls. But, remember, you get what you pay for. A better solution would be some sort of *virtual remote control* (VRC), which would use one of the most powerful home-entertainment controllers ever designed—your computer.

And by powerful multimedia-capable computer, I mean, of course, a Mac. Forget a Windows®-based PC, if you want a reliable, fast, intuitive, viral-free computer. Hey, I'm not some one-trick pony, either. I've been around the personal computer world. My repertoire reads like a PC *Who's Who* listing, including names like Amiga, Apple, Atari, Commodore, Heath,

Robo, Can You Hear Me?

There are 66 Robosapien command key functions that can be accessed via its remote control (function #21 is really a shift key used for stacking two additional functions onto each key). A visual cue response is generated by the Robosapien LED eyes for assisting you with debugging some of the command key functions (Fig. 2-2). Only a small subset of these functions was used in this hack.

1. Right Arm Up
2. Right Arm Down
3. Right Arm In
4. Right Arm Out
5. Tilt Body Right
6. Left Arm Up
7. Left Arm Down
8. Left Arm In
9. Left Arm Out
10. Tilt Body Right
11. Turn Right
12. Walk Forward
13. STOP Button
14. Turn Left
15. Walk Backward
16. (R>) Right Sensor Program
17. (S>) Sonic Program
18. (L>) Left Sensor Program
19. (P>) Program Play
20. (P) Master Command Program
21. (SELECT: Shift for Green Keys & Orange Keys)

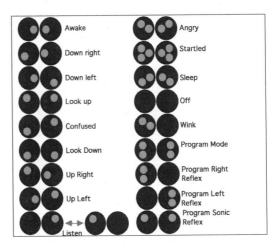

FIGURE 2-2 Yes, you can read Robosapien's mind...just look into its eyes. Emotions and system status events are depicted by various LED lighting combinations. (Photograph © WowWee Ltd.)

22. Right Hand Thump
23. Left Hand Pickup
24. Lean Backward
25. Right Hand Throw
26. Sleep
27. Left Hand Thump
28. Left Hand Pickup
29. Lean Forward
30. Left Hand Throw
31. Listen
32. Forward Step
33. Right Turn Step
34. Backward Step
35. Right Sensor Program Execute
36. Master Command Program Execute
37. Wake Up
38. Reset
39. Left Turn Step
40. Left Sensor Program Execute
41. Sonic Sensor Program Execute
42. Right Hand Sweep
43. High 5
44. Right Hand Strike 1
45. Burp

46. Right Hand Strike 2
47. Left Hand Sweep
48. Talk Back
49. Left Hand Strike 1
50. Whistle
51. Left Hand Strike 2
52. Bulldozer
53. Right Hand Strike 3
54. Oops!
55. Demo1
56. All Demo
57. Power Off
58. Roar
59. Left Hand Strike 3
60. (Select) Return to RED Command Functions
61. Demo2
62. Dance Demo
63. <, < Combination "Right Walk Turn"
64. >, > Combination "Left Walk Turn"
65. Forward, Forward Combination "Slow Walk Forward"
66. Backward, Backward Combination "Slow Walk Backward"

IBM, Microsoft, Sun, and Timex/Sinclair. Oh, and I was right there in the delivery room during the birth of each of these systems, so I've seen them grow and, in some cases, die. I'm a veritable living cradle-to-grave robber. And nothing gripes me more than seeing some poor soul trying to use a PC for anything other than a doorstop.

If you create, design, build, and program robots on a PC, then you're not even coming close to your hacker potential. A PC will actually stifle your creativity. Its restrictions will strangle your inspiration. You will actually compromise on the most important elements of your project and throw your baby robot out the window with the bathwater. Finally, don't subscribe to the drivel that some vendors try to dump on your shoulders, that their product is "*only* PC-compatible." That's horsefeathers. I have yet to find any worthwhile product that *isn't* available for the Mac, including the immensely popular Parallax BASIC Stamp® (e.g., MacBS2) and LEGO® RCX (e.g., MacNQC).

Turning your Mac into a powerful learning remote control, however, could be a task that would challenge even the most gifted Apple engineer, that is, if it weren't for the hardware/software duo of IRTrans (www.irtrans.com) (Fig. 2-3) and iRed (www.filewell.com/iRed/) (Fig. 2-4). These two finely made Ger-

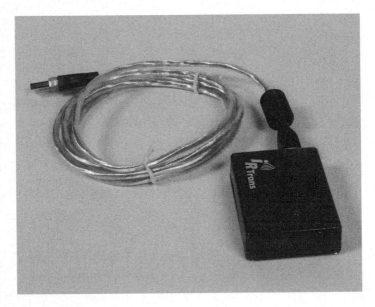

FIGURE 2-3
IRTrans connects to your computer (either Mac or PC) via a USB port. IRTrans can both receive and transmit coded IR signals.

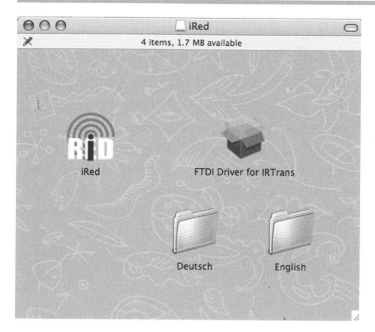

FIGURE 2-4 iRed is the heart of your new Mac IR remote control system. A single FTDI driver must be installed before IRTrans can be connected to your Mac.

man products (designed by the incredibly knowledgeable and helpful Marcus Müller and Robert Fischer, respectively) are perfectly suited for turning your Mac into the ultimate remote control.

Pity the plight of the poor PC IR programmer. Lacking the elegant iRed interface, Windows users must rely on two powerful but complex programs: Girder and MainLobby. Hampered with finicky DLL drivers, an arcane command structure, and the occasional BSoD (Blue Screen of Death), PC-based IR remote controls don't hold a candle to a Mac virtual remote control.

If you find my discourse on the Mac's unchallenged superiority uncomfortable, let me assure you that IRTrans can also be purchased for usage on your PC. This option costs slightly more (€99, or approximately $131), but you also receive a fully licensed version of Girder, Windows and Linux USB drivers, and drivers for Homeseer and myHTPC. Furthermore, you should be able to adapt your PC IR development to the following iRed instructions.

As a demonstration of the ease with which a VRC can be built, let's use IRTrans and iRed for controlling a Robosapien. This is your first Robosapien hack.

IRTrans & iRed

You can order one IR transmitter/receiver bundled with iRed (IRT-USB-Mac) for €85.50 plus a shipping fee of €20.00 from the IRTrans Web site (www.irtrans.com). Additionally, external IR transmitters for 1, 2, 4, and 6 devices (costing €8.00, €14.00, €25.00, and €32.00, respectively) are available to extend the range of the IRTrans (Fig. 2-5). Either an international wire transfer or PayPal can be used for the transaction. You should expect 7 to 10 days for delivery of your IRTrans via Deutsche Post. For more information, contact IRTrans (marcus@irtrans.de) and iRed (ired@filewell.com).5

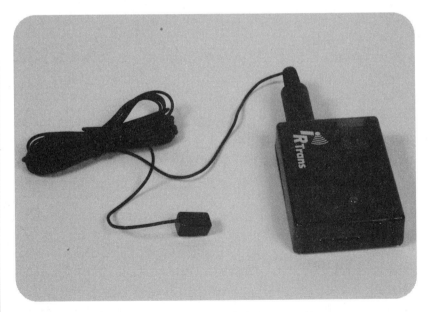

FIGURE 2-5 External IR transmitters can be added to IRTrans for extending its range around corners and into other rooms.

The Hack

STEP 1: INSTALL IRED. Download the iRed application from the iRed Web site (www.filewell.com/iRed/). Expand and unpack the application. You can copy iRed and its English support folder to any location that you want inside your Applications folder. A separate installer program loads the IRTrans FTDI hardware driver. Your Mac must be restarted after this installation.

STEP 2: CONNECT IRTRANS TO YOUR MAC. You will need one free USB port on your Mac and a USB A/B cable. Just plug the cable into IRTrans and then into your USB port. No external power supply is needed; power is drawn from the Mac's USB port.

STEP 3. MAKE IT LEGAL. When you purchase the Mac version of IRTrans, Robert Fischer will e-mail you the license key for iRed. Enter this key into the license panel, and you're ready to create a VRC for Robosapien.

STEP 4: BEGIN A BLANK VRC. After you license iRed, you begin building your new VRC. A blank, brushed aluminum remote control is visually populated with various buttons and text labels via the Edit button (Fig. 2-6). Pressing the Edit button enlarges the template and displays a rudimentary button layout tool and several simple detail commands.

STEP 5: ADD A BUTTON. A button begins life as a selection from the layout tool (Fig. 2-7). Each tool in this layout consists of a row on the final VRC. You can choose from one to four button rows, an empty row, or a text row. To add one of these row layouts to your VRC, just click on the desired tool. For example, to add my first button to the Robosapien VRC, I clicked on the one-button row tool. This added a single button to the top row of the VRC. Then I used the

FIGURE 2-6 **A blank VRC template is the palette that you will use for building your new Robosapien remote control.**

details commands for adding a title to my button. In this case, the letter P represents the master program command on Robosapien (Fig. 2-8).

FIGURE 2-7
When you press the Edit button on the VRC template, the panel expands and a set of button description fields and layout tools become active.

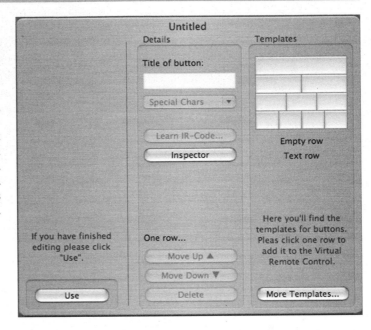

FIGURE 2-8 Your first button. Just select the layout, enter a descriptive label, and use the Learn IR-Code... button for beaming the required IR code from the Robosapien remote control into iRed.

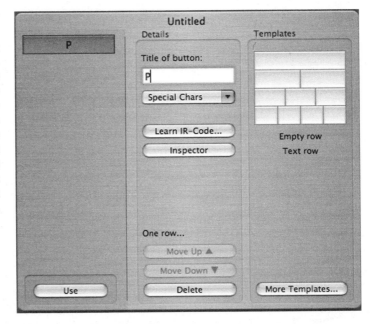

STEP 6: LEARN THE IR CODE. You breathe life into your new button by supplying it with an IR code. Press the Learn IR-Code... to display the Learn IR Code panel (Fig. 2-9). There are three fields on this panel that you will use with your Robosapien VRC. While the button name from the VRC is pre-entered into the description field, you will have to manually enter a remote control name (after you save the VRC, this field will be automatically filled in with the name of your Robosapien VRC). The preselected 39-kHz frequency will work great with the Robosapien remote control. Now get the Robosapien remote control ready for flashing the correct IR signal at the IRTrans. Press the Learn... button and you have 5 seconds for transmitting the IR code to IRTrans. In this case, the Master Program button was pressed for association with the P button on our VRC. If everything goes according to plan, iRed will display an "IR-Code was learned" message, which indicates that the proper IR command has been assigned to the VRC button. You can test your recording by pointing IRTrans at Robosapien and pressing the Test button. If you get the correct robot response, save this button's code and proceed to the next button.

FIGURE 2-9
Just click the Learn... button, and you have 5 seconds to beam a good IR signal from the Robosapien remote control to iRed.

A Lighter Shade of IR

In addition to the common household IR remote control standard, there is another popular format for wireless photon-based communication. IrDA is a high-bandwidth, bidirectional protocol that was a common fixture on older PowerMacs and early PowerBooks. This port was designed for communication with similarly equipped printers, such as the Hewlett-Packard LaserJet 5MP. Wirelessly transmitting printer data at a sizzling 115K bps, the IrDA port was great for line-of-sight (LOS) distances up to about 3 feet and inside a narrow 30-degree-angle optimal operation window. Unfortunately, IrDA ports are not compatible with IR remote controls.

R There N E Other IR Bots?

Once you master your Robosapien with a VRC, you might want to turn your sights onto another IR-controlled robot. For example, the Boe-Bot™ by Parallax (www.parallax.com). Currently being sold in most RadioShack® stores, the Boe-Bot is a two-wheeled robot with several sensor circuits, two servos, and, most important, a programmable microprocessor brain. This brain is a BASIC Stamp 2 microcontroller that is the brainchild of Parallax. Until recently, control of the Boe-Bot was exercised exclusively through PBASIC programming. An IR Remote AppKit from Parallax enables you to control your Boe-Bot through a universal remote control. Using the same IR signal encoding protocol as a Sony® TV, this remote control can be easily configured with the dynamic duo IRTrans/iRed for controlling the Boe-Bot. (*Note:* You will need the BASIC Stamp programming application MacBS2 by the inventive Murat N Konar— www.muratnkonar.com/otherstuff/macbs2/index.shtml—for loading your IR control programs onto the BASIC Stamp.)

STEP 7: BUTTON IT UP. Trying to add IR codes for more than a couple of buttons can become tedious. Luckily, there is a "power user" function built into iRed. Before you can use this function, however, you must first add all your buttons to the VRC. Layout isn't really important, but a properly labeled button will save you some headaches during future HTML and/or AppleScript programming. I elected to name my buttons according to the Robosapien programming language naming convention (i.e., all uppercase letters without any spaces) (Fig. 2-10).

Once you've named your buttons, save your VRC. Be sure to use the same file name that you entered for the Remote Control field on the Learn IR Code panel.

Now, as you add IR codes to each button, both the Remote Control field and the button Description field will be automatically filled in for you. To become an iRed power user, however, you select a button, shift-click on the Learn IR-Code... button, and send the proper IR signal to IRTrans. (*Note:* This power user shortcut bypasses the display of the Learn IR-Code... panel, which is where your time savings comes from.) With just a little amount of dexterity, you can add IR codes to all of the Robosapien buttons very quickly. And with more than 60 Robosapien commands, you will appreciate this power user shortcut.

FIGURE 2-10 **The finished Robosapien VRC. You can use it just like the original Robosapien remote control, but if you really want to build a better IR mouse-trap, then you really have to open it up.**

Ouch, Something's Wrong

TESTING, TESTING. After you've learned IR codes for every button, click on the Inspector button. An Inspector panel enables you to see a list of all Robosapien buttons, review the IR code for each button, and optionally add AppleScript to any of these buttons (Fig. 2-11). There is also a Test tool for sending a button's IR command to IRTrans for transmission to Robosapien. You should test each button's action to ensure that the proper command is sent to Robosapien. You know, if you press the button labeled "DANCE," then Robosapien should dance. Right? This is a great sanity check for making sure that your future multistep master programs will function properly.

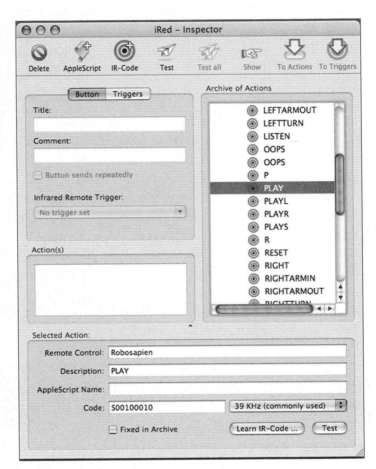

FIGURE 2-11
Use the iRed Inspector tool for verifying that every button has an IR code associated with it.

Rosebud..., One More Thing

SKIN THAT SUCKER. Now it's time to put a pretty face on this VRC. Sure, the brushed aluminum VRC looks OK, but you want a remote control that looks, well, like Robosapien. Built into iRed is a nifty HTML code interpreter that enables you to do just that—skin your VRC. This incredibly easy, but decep-

Girder Guidelines

If your inner masochistic self insists on using Windows for your IR control functions, then you should strongly consider using Girder as your software interface with IRTrans. First of all, though, be advised that Promixis is now responsible for the maintenance, upgrade, and support of Girder. Also, you can purchase Girder for around $20 directly from Promixis' Web site.

Integrating IRTrans with Girder is *very* similar to the procedure used with iRed, only a little more confusing. For example, to setup the IRTrans plug-in within Girder, you must set the address of the IRTrans server. Huh? This entry is the host name, and, luckily, the name *localhost* can be used *if* both the IRTrans client and server are running on the same machine.

Just like with iRed, you must now learn all of your Robosapien IR commands with Girder. After you get all your IR commands entered into Girder, you can use Girder to send commands to Robosapien. Before you can send these IR commands, however, there are two fields that must be filled in for each command. A remote name and a command name must be supplied for each IR command. Both of these entries are fairly straightforward: Remote is the name of the physical remote control device (e.g., Robosapien), and command is the name of the IR command keystroke (e.g., Dance or DANCE). In a nifty shortcut, these two fields can be automatically filled in by just pressing the corresponding key on the physical remote control device, *if* you have already learned that particular key's IR command.

tively powerful, capability requires only four ingredients: a terrific JPEG image file (this is your skin, so let your creative juices go) (Fig. 2-12), image map coordinates for your skin's buttons, an HTML-RC file (you get a template for building this file with iRed), and the new iRed URL HREF scheme (you know, this is the part of an HTML link that defines the type of protocol for accessing the linked document, e.g., http: or ftp:; in this case ired: is the new scheme). The latter ingredient is added with this type of HTML syntax: <a href="ired://send?Robosapien&command">.

FIGURE 2-12
A skinned Robosapien VRC. Just open a robosapien. html file (see Listing 2-1) inside iRed and you can play the six big Robosapien programs that I coded into my VRC.

Note: You replace the send parameter with call for accessing iRed Apple-Script programs.

Consider this simple HTML file, which controls a six-button Robosapien image map (Listing 2-1). Each button streams a large multistep program to IRTrans, which, in turn, transmits the required IR signals to Robosapien. The power of this command concatenation technique is impressive to witness as each signal is acknowledged by Robosapien and then automatically executed.

Listing 2-1

```
<!DOCTYPE html PUBLIC "-//W3C//DTD HTML 4.01 Transitional//EN">
<!—<ired>
/* comments allowed!! */
/* add 24 to height of table etc. */
width=498;
height=450;
title="Robosapien";
</ired>—>
<html>
    <head>
        <meta http-equiv="content-type" content="text/html;charset=iso-8859-1">
        <title>Robosapien</title>
    </head>
    <body bgcolor="#ffffff" leftmargin="0" marginheight="0" marginwidth="0" topmargin="0">
        <table border="0" cellspacing="0" cellpadding="0"><tr height="426">
        <td width="498" height="426">
        <img src="robosapien.jpg" alt="VRC" width="498" height="426" usemap="#robomap" border="0">
        <map name="robomap">
            <area shape="rect" coords="303,91,388,177"
href="ired://send?Robosapien&P&Robosapien&HIGH5&Robosapien&BACKWARD&Robosapien&TALK
BACK&Robosapien&BACKWARD&Robosapien&RIGHTTURN&Robosapien&STOP&Robosapien&L&Robo
sapien&BACKWARD&Robosapien&DANCE&Robosapien&LEFTTURN&Robosapien&WHISTLE&Robosapi
en&STOP&Robosapien&FORWARD&Robosapien&PLAY" alt="Party Animal">
            <area shape="rect" coords="324,200,433,299"
href="ired://send?Robosapien&P&Robosapien&OOPS&Robosapien&BURP&Robosapien&OOPS&Robos
apien&BURP&Robosapien&OOPS&Robosapien&BURP&Robosapien&OOPS&Robosapien&BURP&Robos
apien&OOPS&Robosapien&BURP&Robosapien&OOPS&Robosapien&BURP&Robosapien&OOPS&Robos
apien&WHISTLE" alt="Gas Attack">
            <area shape="rect" coords="199,279,298,362"
href="ired://send?Robosapien&R&Robosapien&BACKWARD&Robosapien&LISTEN&Robosapien&LEFT-
TURN&Robosapien&LISTEN&Robosapien&RESET&Robosapien&LISTEN&Robosapien&STOP&Robosapie
n&L&Robosapien&BACKWARD&Robosapien&LISTEN&Robosapien&RIGHTTURN&Robosapien&LIS-
TEN&Robosapien&RESET&Robosapien&LISTEN&Robosapien&STOP&Robosapien&RESET&Robosapien&
FORWARD" alt="Wander About">
            <area shape="rect" coords="70,200,166,295"
href="ired://send?Robosapien&R&Robosapien&OOPS&Robosapien&RESET&Robosapien&P&Robosapi
en&STOP&Robosapien&L&Robosapien&BURP&Robosapien&RESET&Robosapien&P&Robosapien&STO
P&Robosapien&RESET&Robosapien&LEFTARMIN&Robosapien&RIGHTARMOUT&Robosapien&TILTRIGH
T&Robosapien&RIGHTARMIN&Robosapien&RIGHTARMOUT&Robosapien&R&Robosapien&RIGHTARMI
N&Robosapien&LEFTARMOUT&Robosapien&TILTLEFT&Robosapien&LEFTARMIN&Robosapien&LEFT-
ARMOUT&Robosapien&L&Robosapien&HIGH5&Robosapien&PLAY" alt="Pull My Finger">
            <area shape="rect" coords="106,82,194,177"
href="ired://send?Robosapien&P&Robosapien&TALKBACK&Robosapien&PLAYR&Robosapien&PLAYS
&Robosapien&PLAYL&Robosapien&PLAYR&Robosapien&PLAYS&Robosapien&PLAYL&Robosapien&TA
LKBACK&Robosapien&PLAYR&Robosapien&PLAYL&Robosapien&PLAYR&Robosapien&PLAYL&Robosa
pien&PLAYR&Robosapien&BURP&Robosapien&OOPS&Robosapien&R&Robosapien&TILTLEFT&Robosap
ien&LEANFWD&Robosapien&TILTRIGHT&Robosapien&LEANBWD&Robosapien&TILTLEFT&Robosapien
&TILTRIGHT&Robosapien&STOP&Robosapien&S&Robosapien&LEANBWD&Robosapien&LEAN-
FWD&Robosapien&LEANBWD&Robosapien&LEANFWD&Robosapien&LEFTARMOUT&Robosapien&RIG
HTARMOUT&Robosapien&STOP&Robosapien&L&Robosapien&LEANFWD&Robosapien&TILTRIGHT&Ro
bosapien&LEANBWD&Robosapien&TITLLEFT&Robosapien&RIGHTARMOUT&Robosapien&LEFTAR-
MOUT&Robosapien&STOP&Robosapien&PLAY" alt="Song + Dance">
            <area shape="rect" coords="205,170,298,258"
href="ired://send?Robosapien&S&Robosapien&TALKBACK&Robosapien&STEP&Robosapien&STEP&R
obosapien&LEFTTURN&Robosapien&RIGHTTURN&Robosapien&BACKWARDS&Robosapien&STOP&Ro
bosapien&PLAY" alt="Sentry Duty">
        </map>
        </td></tr></table>
</body>
</html>
```

Oh Yeah? Make Me

Remarkably, there are four different modes that can be programmed on the Robosapien; one master program and three independent sensors. Wow, or should I say, WowWee!

1. Master Program (P): general control programs.

2. Right Sensor Program (R>): Robosapien's right-side finger, toe, or heel touch sensor.

3. Left Sensor Program (L>): Robosapien's left-side finger, toe, or heel touch sensor.

4. Sonic Sensor Program (S>): a loud sound or body hit sensor on its body.

ONE MORE STEP. The iRed software is even more powerful than this short Robosapien remote control demonstrates. For example, a standard battery-powered remote control (available from Marcus Müller at IRTrans) can be used for sending signals to your Mac, which, in turn, can then be used for controlling other IR devices, as well as other Mac apps. All of this magic is made possible by inserting AppleScript into iRed. Therefore, apps like iTunes can be controlled via an external remote control, which triggers IRTrans to transmit coded IR signals to your amplifier for turning it on and setting volume, balance, treble, bass, and loudness via a VRC. Finally, Airport Express sends music wirelessly to your Mac IR-controlled home stereo. Now that's the ultimate remote control.

CHAPTER 3

Pump You Up

My God, it seems like magic—a product that produces electricity just by sitting in the sun. This isn't some bit of electronic legerdemain, however. This is a photovoltaic (PV) cell.

Although some of us old dogs might refer to these marvels of modern technology as "solar cells," they are, in fact, more accurately called photovoltaic cells (Fig. 3-1). In a nutshell, a photovoltaic cell is capable of converting sunlight energy into electrical energy. Or, more technically, sunlight photons strike the photovoltaic cell and discharge electrons that gather to form an electric current. Today you see these little amazing power plants everywhere: from calculators to outdoor lighting systems.

This miracle began in 1839, when Edmund Becquerel discovered that when sunlight was absorbed by certain materials, it could generate both heat and a small amount of electricity. Usage of this discovery was largely confined to measuring light levels in photography. Then in the mid-1950s, some radical improvements in manufacturing processes, fueled by the fledgling NASA space exploration programs, catapulted Becquerel's discovery into the birth of today's photovoltaic cell (Fig. 3-2).

As previously mentioned, photovoltaic cells convert sunlight into electricity with nary a moving part, consumable fuel source, or battery connection. Typically, PV cells are made of chemically refined silicon—a semiconductor.

FIGURE 3-1 The solar cell is the basic building block of a photovoltaic system. Individual cells can vary in size from about 1 centimeter (½ inch) to about 10 centimeters (4 inches) across. This crystalline silicon cell was manufactured by Shell Solar Industries. (Photograph courtesy of Shell Solar; Siemens Solar Industries became Shell Solar in April 2002)

FIGURE 3-2 After decades of use on Earth and in space, a photovoltaic cell saw its first use on another planet in 1997 when Sojourner began to explore Mars. High-efficiency PV cells mounted on top of the Sojourner vehicle provided 16 W of power at noon on Mars, which was enough to perform the mission of the day. (Photograph courtesy of NASA and NSSDC Photo Gallery)

Either boron or phosphorous are mixed with the silicon to enhance the PV cell's ability to release electrons when tickled by photons. Selenium, gallium-arsenide, and cadmium telluride have been substituted for silicon in the manufacturing of PV cells with varying results.

Following the release of these electrons inside the PV cell, the building electrical current is funneled into a set of attached wires. Additional cells can then be wired together for generating more current and higher voltage. Groups of PV cells can then be arranged side by side into a rectangular shape called a module. Likewise, several modules can be then tied together forming an array.

When PV arrays get big enough for some serious power generation, they are generally mounted on the roof of a building angled south for an optimal dose of daily sunshine. Some more sophisticated PV arrays are mounted on moveable tracking systems that follow the sun's path for prolonged maximum solar exposure.

One of the most amazing advancements in solar-powered arrays has been the recent development of low-cost, high-yield, flexible, thin-film PV cells (Fig. 3-3). Fabricated on a tough, resilient, flexible plastic polymer substrate (some even have a fabric backing for added strength), these thin-film PV cells are able to generate voltages roughly equivalent to their rigid counterparts. A vast improvement over the rigid PV cells, however, these thin-film arrays are typ-

FIGURE 3-3 Professor Bernard Kippelen holding an electronic circuit fabricated on a flexible plastic substrate. The solar cells he is developing can be integrated with such organic electronic devices. (Photograph courtesy of School of Electrical and Computer Engineering at the Georgia Institute of Technology)

Tower of Power

What do you call a 3,280-foot-tall tower that can generate 200 megawatts of power? Lacking any real inspiration here, EnviroMission Limited calls its electric dynamo the—are you ready for this—Solar Tower. Construction of this remarkable solar power plant is scheduled to begin in 2006.

Situated on a 25,000-acre sheep farm near Mildura, Victoria, Australia, and costing EnviroMission Limited a cool $1 million, the future Solar Tower consists of three main components: a 25,000-acre collector zone (dubbed the greenhouse), the 3,000+-foot-tall chimney tower, and 32 wind turbines that are powered by the convection currents rising from the greenhouse through the tower.

As fantastic as this concept sounds, the Solar Tower technology has been successfully tested on a dwarf version pilot plant in Manzanares, Spain. Operated for seven years from 1982 to 1989, this Solar Tower "mini" generated 50 kilowatts (kW) of green energy throughout its existence.

Based on the theoretical power projections of the Solar Tower, there will be enough power to operate 200,000 typical Australian homes, while, in the words of EnviroMission, abating "900,000 tonnes of greenhouse producing gases from entering the environment annually" (www.enviromission.com.au).

ically supplied in sheets or rolls, can be connected together, and are weather-resistant. Based on its ease of operation and fabric-like qualities, thin-film PV cells can be easily integrated into portable light-weight products. Even the U.S. Army has begun to notice thin-film PV arrays, and studies are currently underway regarding the incorporation of these flexible power supplies into tents, backpacks, and even uniforms.

Several years ago I felt that thin-film PV cells could be installed on an aircraft as its sole means of fuel. Using a plan derived from a 1913 Etrich Taube, I set to work devising a "Solar Dove." My first solar hack. Regrettably, my Solar Dove never saw the light of day. The same can't be said about your second Robosapien hack. Unless you have stock in one of the major alkaline battery manufacturers, rechargeable batteries are your best choice for keeping Robosapien dancing to its own funky beat.

The Hack

STEP 1: OK, YOU'VE GOT TO START SOMEWHERE. Yeah, I know; this sounds like a no-brainer, but you've got to remove your alkaline batteries before you can use rechargeable batteries.

STEP 2: CHARGE IT. The particular model of solar battery charger that I used can only accept two D-size batteries at a time (Fig. 3-5). Likewise, the intensity of sunlight can dramatically affect the battery charge time. So strive for maximum sunlight and give yourself about 9 to 18 hours for each pair of batteries. A nifty meter is built into the solar battery charger sold by C. Crane (Fig. 3-6). This meter will help you estimate your charging time. For example, a solar intensity of 120 milliamperes (mA) will

take approximately 12 hours to charge. To obtain the maximum charging capability, you can use the built-in prop stand on the bottom of the charger (Fig. 3-7). This stand will help you achieve the best angle for keeping your charger's best face toward the sun (Fig. 3-8).

FIGURE 3-5
Hey, no power cords; no kidding. Solar power, baby. Free Robosapien power.

FIGURE 3-6
The built-in meter helps you maintain the maximum sunlight strength for optimal charging.

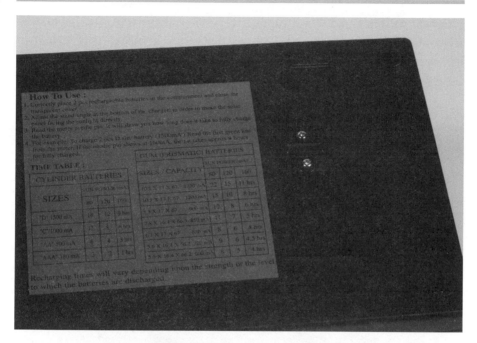

FIGURE 3-7
**A convenient
prop stand is
on the back for
angling the
charger toward
the sun.**

FIGURE 3-8 **Keep
the charger's
best face
towards the
sun—its PV
array face.**

STEP 3: PLUG 'N GO. Once your batteries are fully charged (this operation takes me about two days of full Mississippi sunlight) (Fig. 3-9), just turn Robosapien over and install your charged rechargeable batteries (Fig. 3-10). Enjoy your free power.

FIGURE 3-9
You can only charge two batteries at a time. So plan ahead for your charging times.

FIGURE 3-10
Install the rechargeable batteries in Robosapien.

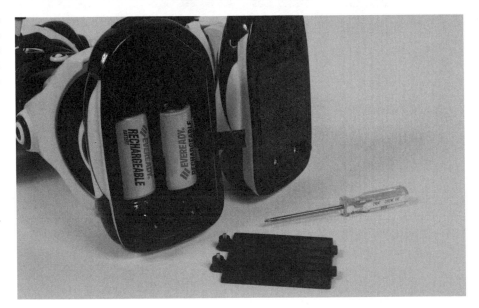

Ouch, Something's Wrong

The most common fault with these solar battery chargers is inadequate sunlight intensity. For example, you hackers in the north (e.g., North Dakota) will need a greater length of time to charge your batteries. Don't be alarmed if your charger's meter indicates a solar power current of 80 mA or less. Although hackers along the Gulf Coast can safely down a mint julep (or three) waiting for all four batteries to charge. In this case, I have easily obtained solar power current readings of 160 mA.

If this lengthy power charging time commitment really bugs you, consider buying two solar battery chargers. Then you will get your Robosapien juiced up in half the time. Also, you should consider adding rechargeable batteries to your other Robosapien goodies. For example, Mini Robosapien (Figs. 3-11

FIGURE 3-12
The Mini
Robosapien
batteries are
ready to go in
one charging
session.

FIGURE 3-13 A
rechargeable
9-V battery
can be added
to the IR
remote
control.

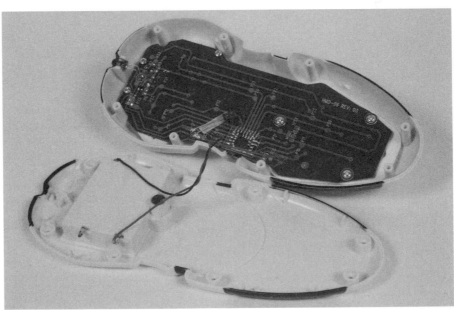

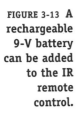

and 3-12) and the Robosapien IR remote control (Fig. 3-13) are perfect candidates for receiving rechargeable batteries. Your local landfill will thank you (Fig. 3-14).

FIGURE 3-14
The plastic cup that comes with Robosapien is ideal for holding batteries prior to charging.

Million Solar Roof Initiative

In June 1997, President Bill Clinton announced the Million Solar Roof Initiative. This initiative assists businesses and individuals in installing solar energy systems on 1 million buildings across the United States by 2010. A partnership among the U.S. Department of Energy, state and local governments, solar industry manufacturers, and electricity service providers strengthens the demand for solar technologies and streamlines the installation process. There are three key features in this initiative:

1. Improve financing options for solar installations.
2. Encouraging builders and developers to include solar energy systems in new construction.
3. Continuing to develop and improve on solar technologies.

Have you seen any involvement in your local community with the Million Solar Roof Initiative? No; then find out what you can do to make this dream a reality at www.millionsolarroofs.org.

Rosebud…, One More Thing

OK, so this hack didn't really test your abilities, huh? Well, how about installing the battery charger inside Robosapien? Don't even think about attempting this hack if you aren't extremely familiar with power supplies and heat sinks. One incorrect connection, and you could melt the Robosapien plastic, explode the D cell batteries, or both. Throwing caution to the wind, however, this hack can make an extremely clean design.

The battery charger circuit is based on the LM317 adjustable, three-terminal, positive regulator analog integrated circuit (IC) (Figs. 3-15 and 3-16). With an output current of more than 1.5 amperes (A) along with an easily programmable output voltage, the LM317 is ideal for cramming a battery charger circuit into a tight area. Even better, you will only need one resistor (22 ohm) for setting the output voltage for charging the NiCads.

Derived from a National Semiconductor (www.national.com) data sheet application hint (e.g., LM117/LM317A/LM317), this circuit will generate an

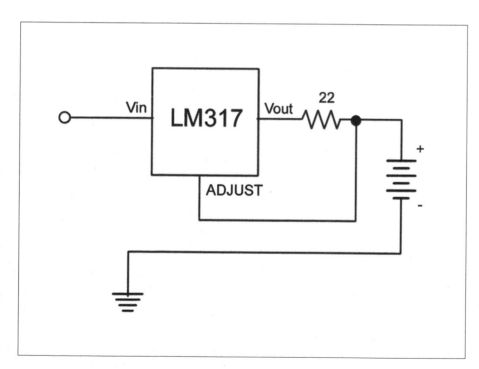

FIGURE 3-15
The schematic diagram for a LM317 battery charging circuit.

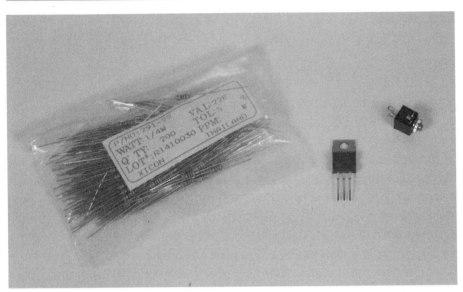

FIGURE 3-16 Only three components are needed for building one LM317 battery charging circuit. You can make one charging circuit for each Robosapien foot, or, by wiring both battery compartments together, you can use one circuit for charging all four batteries.

output charging current of 43 to 50 mA. The input supply voltage should be around +12 volts (V). This value can vary *slightly* without adverse effect; just don't exceed +20 V (or drop lower than +9 V). Use any one of the commonly available wall power adapters (i.e., wall wart) for your supply, preferably with a 3.5 millimeter (mm), 2.5 mm, or barrel plug. Now mount your circuit under the battery pack inside each foot sole, and drill a hole into the back heel or tendon area of the foot for accommodating a corresponding power supply input receptacle. Wire it up, check it with your voltmeter, and you're ready to charge your batteries in situ. Just remember to unplug the power supply before you take Robosapien for a stroll.

CHAPTER 4

All of the Best Robots Have Digital Eyes

Do you have an AirPort installed in your home or office? If not, then you are truly missing out on one of the greatest advancements of the 20th century—wireless computing. Yes, way back in 1999, Apple Computer helped establish the 802.11 wireless networking standard in personal computers, and subsequent improvements resulted in the adoption by Apple of the 802.11g high-speed wireless technology into its AirPort Extreme. At data rates of approximately 54 megabits per second (Mbps), the AirPort Extreme can handle up to 50 users sharing a single Internet connection, as well as granting them access into an Ethernet network and allowing file sharing. Oh, and for you diehard PC fans, AirPort Extreme plays well with Windows, too.

So what's with all of these standards, anyway? Exactly what is 802.11 and 802.11g and Wi-Fi and Bluetooth® wireless technology, for that matter? Our saga begins with the Institute of Electrical and Electronics Engineers (IEEE) creating a wireless Ethernet (802.3) in 1990. Officially named IEEE 802.11, this wireless Ethernet utilized radio frequencies on the 2.4-GHz band for building wireless local area networks (WLANs). Consumers were finally able to benefit from this new standard in 1999, when Apple introduced the AirPort wireless

It All Started Back in the 10th Century

Now what's the deal with this name—Bluetooth? According to the Bluetooth Special Interest Group (SIG) Web site (www.bluetooth.org), a 10th-century Danish king named Harald Blatand, or Harold Bluetooth, was selected as the code name for this new wireless technology by the Bluetooth Trade Association. King Bluetooth was a unifier in Scandinavia, and the members of the Trade Association felt that his name was apt for inspiring collaboration among diverse industries.

You can also sense this Scandinavian origin in the Bluetooth wireless technology logo (Fig. 4-1). If you look closely you can see the runic letters H and B superimposed over each other. Oh, it might help you to know that a runic H looks a lot like an asterisk (*).

FIGURE 4-1 The Bluetooth word mark and logos are owned by the Bluetooth SIG, Inc.

network system, 802.11b. Now an easy and affordable solution, users could finally "sever the tether" in office, school, and home networks.

Capable of delivering data up to 11 Mbps, the 802.11b AirPort was replaced in 2003 by the 802.11g-compliant 54 Mbps AirPort Extreme. Luckily, AirPort Extreme is backward compatible with older 802.11b appliances. Therefore, AirPort Extreme products will integrate seamlessly with any Wi-Fi Certified network—both 802.11b and 802.11g.

Wi-Fi Certified or Wireless Fidelity Certified is administered by the nonprofit Wi-Fi Alliance. This certification indicates that a product has been tested with other wireless products and has been found to be interoperable with other 802.11 devices. So, *any* product that is Wi-Fi Certified will work with *any* other Wi-Fi Certified product. All is not bliss in the wireless world, however. There is an odd man out.

As a result of the popularity of the 2.4-GHz band, 802.11g networks can occasionally receive interference from other devices like microwave ovens and cordless telephones. Attempting to nip this trouble in the bud, the 802.11a standard was released. Sounds odd, doesn't it? A new standard with a name preceding an older existing standard. So be it, the 802.11a standard delivers its data on the 5-GHz radio band at a rate up to 54 Mbps. Systems using 802.11a also take a power hit of 2W to 2.5W compared with 1W to 1.5W for 802.11b and 802.11g, respectively. Unfortunately, you guessed it, 802.11a devices aren't compatible with 802.11b networks. Furthermore, some countries haven't certified 802.11a. So this standard has been left waiting at the alter.

Well, what about this Bluetooth thing? The official statement from the Bluetooth SIG states that: Bluetooth wireless technology is set to revolutionize the personal connectivity market by providing freedom from wired con-

nections—enabling links and providing connectivity between mobile computers, mobile phones, portable handheld devices, and much more. Bluetooth wireless technology redefines the very way we experience connectivity. The Bluetooth SIG, comprising leaders in the telecommunications, computing, consumer electronic, network and other industries, is driving the development of the technology and bringing it to market. The Bluetooth SIG includes Promoter companies 3Com, Agere, Ericsson, IBM, Intel, Microsoft, Motorola, Nokia, Toshiba, and more than 2,000 Associate and Adopter companies. End quote.

On the technological side of the house, however, the Bluetooth specification is a low-cost, low-power radio standard that is used for connecting devices. Global acceptance of this specification has helped incorporate limited range wireless communication into everything from automobiles to ZigBee™. Based on a nifty time-sharing architecture featuring frequency hopping and tiny packet sizes, Bluetooth uses the 2.4-GHz radio band with a range of approximately 30 feet.

Leading the pack, again, Apple PowerBook G4 portables are the first computers to offer Bluetooth 2.0+ enhanced data rate (EDR). Other computers are stuck in the older Bluetooth 1.x support. Bluetooth 2.0+EDR, while backward-compatible with Bluetooth 1.x, is up to three times as fast as the older standard. A maximum data rate of up to 3 Mbps is possible with Bluetooth 2.0+EDR. This throughput plus the peripheral nature of the connectivity feature has enabled some vendors to describe Bluetooth as "wireless USB."

This wireless standard propagation stuff is far from over. There's even a whiff in the air that yet another standard is being proposed. The Intel-sponsored 802.11n standard will up the ante for data rates to 200+ Mbps. Anyone can see that wireless communication is the future of connectivity.

In my case, I have an office that is totally wireless. By installing an AirPort Extreme with an accessory Dr. Bott ExtendAIR Omni antenna (www.drbott.com), I was able to flood a 2,000-square-foot workspace with 100% wireless access. The accessory antenna greatly extended the range of the AirPort Extreme, where a recent online conference was able to be conducted almost 200 feet from the base station on wirelessly equipped iBooks. Any wireless disbeliever using a battery-powered notebook computer in that kind of a remote situation is quickly convinced that now is time to cut the umbilical—all of them.

Gosh, We Could Sure Use
Another Standard

Just when you thought that it was safe to use wireless technology, along comes another standard. ZigBee, an IEEE 802.15.4 packet data protocol for small-scale wireless networks (less than 250-kbps data rate) specifying Media Access Control layer (MAC) and Physical layer (PHY), uses the 2.4-GHz radio band and is going head to head with Bluetooth wireless technology systems. A 50-year-old company with its roots firmly planted in radio frequency and wireless technology, Freescale Semiconductors is delivering ZigBee technology for home and industry use. A wholly owned subsidiary of Motorola, Freescale Semiconductor has made two primary contributions to ZigBee that are extremely important to robot builders:

1. ZigBee-compatible sensors. Freescale offers acceleration, pressure, and ion and photo smoke sensors. Based on micro-electromechanical systems technology, these sensors can be readily interfaced with standard hardware systems.

2. MAC software. Freescale's MAC software is both 802.15.4 compliant and smaller than a comparable Bluetooth technology. Therefore, less on-chip memory is dedicated to network connections. According to Freescale, 24 kb of RAM is needed for its MAC (www.motorola.com/zigbee).

What better way to bring Robosapien into your brave new world than with a new wireless communication system. Rather than ripping out the established IR control system, this hack will integrate a wireless video camera into the Robosapien. The resulting images can be displayed on your computer, viewed on a monitor, and/or recorded on a VCR. No burglar will be safe when breaking into a karate choppin', video recordin', protected-by-Robosapien household.

The Hack

STEP 1: OFF WITH THE EARS. Saw off the two white ear pivot joints shown in Fig. 4-3 from the Robosapien head (see Fig. 4-4). A plastic model hand saw works great for this step. Alternatively, you can try a pair of flush-cut snips or a power saw. If you elect to use the power saw option, use the slowest speed available for your saw. Too high a speed and the head plastic will melt. My fav tool for this removal has been a screwdriver. Yeah, I know, but it worked, and I didn't mar the plastic.

FIGURE 4-3
To get inside this lug's head, you must first remove the ears.

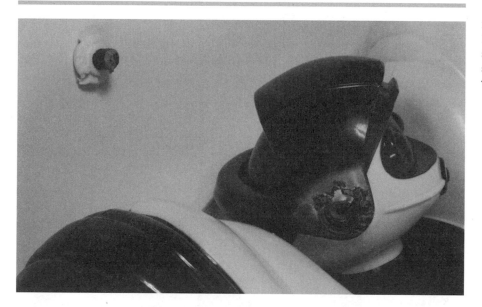

FIGURE 4-4
In a pinch a
screwdriver can
safely pry off
these covers.

STEP 2: LUKE, I'M YOUR FATHER. Carefully remove the Robosapien helmet and the moveable visor (Fig. 4-5). Save these items for reassembly later.

FIGURE 4-5
**Both the
helmet and
the visor are
easily lifted
off of the
Robosapien
head. Save
the visor for
covering the
rear of the
head, later.**

STEP 3: IN A PINCH. Now this step is really tough, but with a little work you can get a near perfect result. Turn Robosapien around so that the back of the head is facing you. You are going to remove the translucent purple plastic from the back of the head (see Fig. 4-6). To remove this plastic, I have found that it is best to pinch it off with a pair of hefty pliers. Watch your fingers here; open the jaws of the pliers wide enough to take a good bite into the purple plastic, and squeeze the heck out of the pliers. After a couple of good healthy squeezes, you should have a mangled pile of purple plastic and a roomy opening in the back of the Robosapien's head (Fig. 4-7).

FIGURE 4-6
Yeah, it's a tough step, but someone's got to do it. This purple skull cap must be removed.

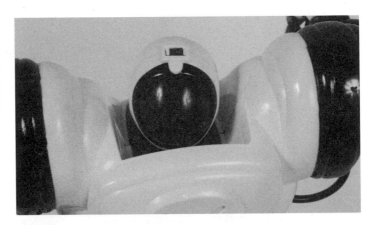

FIGURE 4-7
Success! It ain't pretty, but it worked, and now the real fun can begin.

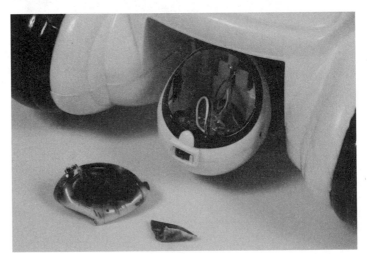

STEP 4: BLINDED BY SCIENCE. Take a deep breath and remove the LED eyes from Robosapien (see Fig. 4-8). You will have to unscrew two screws that hold the LED assembly inside the head. There is one other screw (it was the screw that was actually holding the purple plastic that you just removed earlier, in place) that needs to be removed, also (see Fig. 4-9).

FIGURE 4-8
Removing the LED eyes and the external "bug-eye" plastic cover is a two-step process. First, remove the three screws that hold the LED eyes circuit board in place.

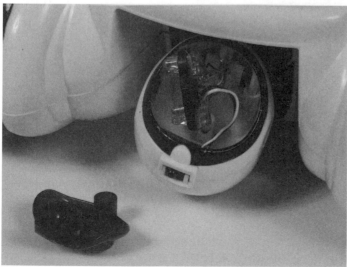

FIGURE 4-9
Second, wiggle the purple plastic "face" that covered the LED eyes out through the front of the Robosapien head.

Bluetooth on a Chip

You might not have heard of them before, but Cambridge Silicon Radio (CSR; www.csr.com) is one of the big players in the single-chip radio device market of short-range wireless communication (Fig. 4-10). The CSR main offering is BlueCore (Fig. 4-11).

CSR's BlueCore3 third-generation Bluetooth silicon for mobile applications

Photo ref: PEM3813   Contact: EML +44 (0)20 8408 8000   email: *csr@eml.com*   EML

FIGURE 4-10 **Cambridge Silicon Radio's new generation of Bluetooth silicon pushes the Bluetooth market. BlueCore3 is the first complete implementation of the Bluetooth v1.2 standard (including all optional features of the standard) which improves coexistence of Bluetooth with other 2.4-GHz systems. (Photograph courtesy of Cambridge Silicon Radio)**

BlueCore is a fully integrated 2.4-GHz radio, baseband, and micro-controller used in more than 60% of all qualified Bluetooth v1.1 and v1.2 enabled products from Apple, Dell, IBM, Motorola, NEC, Nokia, Panasonic, RIM, Sharp, Sony, and Toshiba. Whew!

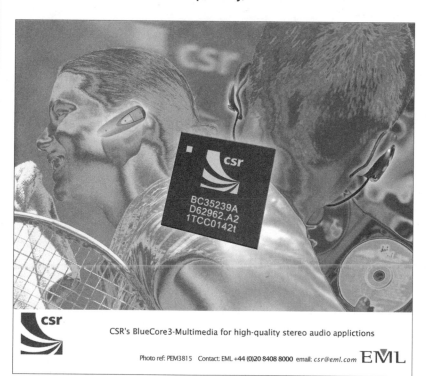

CSR's BlueCore3-Multimedia for high-quality stereo audio applictions

Photo ref: PEM3815   Contact: EML +44 (0)20 8408 8000   email: *csr@eml.com*   EML

FIGURE 4-11 BlueCore3-Multimedia contains an open platform DSP coprocessor that is particularly suitable for enhanced audio applications. The solution includes a 16-bit stereo audio CODEC with dual ADC and DAC for stereo audio. With integrated amplifiers for driving microphone and speakers, BlueCore3-Multimedia is a highly compact Bluetooth solution requiring only minimal external components. BlueCore3-Multimedia is available in LFBGA package. (Photograph courtesy of Cambridge Silicon Radio)

STEP 5: IF I HAD ONE GOOD EYE. Insert the wireless video camera into the void left by the absent LED eyes (Fig. 4-12). Route the camera's antenna and the battery connector backward out of the head (Fig. 4-13). You can hold the camera in place by reattaching the LED circuit board to the back of the head (Fig. 4-14). Carefully turn the LED circuit board around (i.e., so that the LEDs are facing backward out the back of the head) and then rotate it 90

FIGURE 4-12 **Everybody's there, but nobody's home. The empty head is ideal for holding a wireless video camera.**

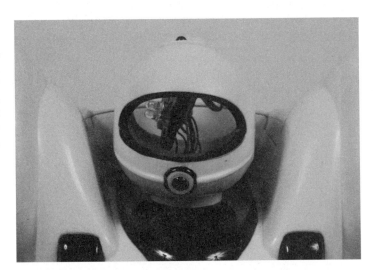

FIGURE 4-13 **It's a perfect fit; both the battery connector and the antenna can conveniently exit the rear of the Robosapien head.**

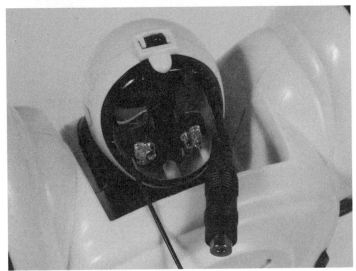

degrees—either clockwise or counterclockwise, your choice. Now fix the circuit board to the back of the head with the last screw that you removed in Step 4 (i.e., the one that was holding the purple plastic in place). Finally, locate some small machine bolts, washers, and nuts for reattaching the visor to the *back* of the head. This will help "clean up" the trepanning that you did on the back of the head. I've used 4-40 hardware for this step and it will work with a steady hand and *lots* of patience (Fig. 4-15).

FIGURE 4-14 **The wireless camera is held in place by the back of the LED eyes circuit board.**

FIGURE 4-15 **The visor that was removed in Step 2 is now reattached backward for hiding the hole in the back of Robosapien's head.**

STEP 6: POWER TO THE PEOPLE. You can keep the 9-V battery either external or internal (Fig. 4-16). If you elect to go internal, bore a small hole into the seam of the front and back chest plates near the neck line. Loosen the screws that hold the two chest plates together, take a reamer and enlarge the hole for accommodating the battery connector cable. Run the 9-V battery connector down into the chest cavity. (*Note:* The reamer is the most frequently used tool for safely hacking the plastic of Robosapien. You can purchase a good reamer for under $6 from Digi-Key.) Make sure that you wrap the 9-V battery with some antistatic foam prior to sealing it up. Enjoy watching RSTV.

FIGURE 4-16
A 9-V battery is all the power that this wireless video camera needs for broadcasting.

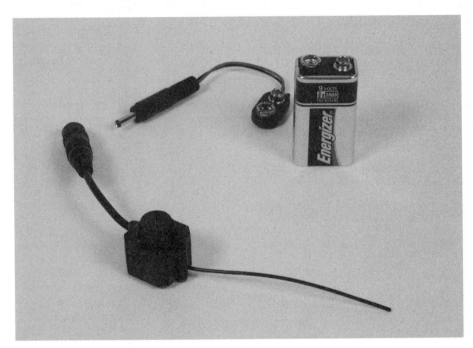

Ouch, Something's Wrong

If Robosapien fails to startup after this hack, your 9-V battery might be shorting the main circuit board. Add some extra antistatic foam around the battery. Another problem can come from the Robosapien motors causing interference with the wireless camera's signal. Try adding some foil around the upper chest cavity motors *and* the two motors that are located in the upper arms. Finally, if the Robosapien IR remote control doesn't appear to be working anymore, you might have damaged the IR detector that is located on the very top of the Robosapien head. This damage can come from jerking around the LED "eyes" circuit board and breaking the detector's delicate wiring or from hitting the detector with one of your hand tools during helmet removal. You could try to salvage this damage by repairing the wiring or replacing the detector, but you might have to skip ahead to Chapter 12 and learn how to add a new microcontroller brain to your now brain-dead bot.

Rosebud..., One More Thing

I love hacks like this one. Pure and simple with a tremendous payback—good bang for your buck. If you're like me, however, the ZigBee technology sounds very intriguing. Luckily, there is a low-cost opportunity for playing around with this new standard.

Freescale Semiconductor makes a great Developer's Starter Kit (DSK) for experimenting with 802.15.4 ZigBee wireless technology. Consisting of two reference design boards equipped with accelerometer sensors, MC13192 2.4-GHz RF transceiver data modems, and MC9S08GT60 microcontrollers (Fig. 4-17), the DSK includes all of the software and documentation (and even a trial version of

FIGURE 4-17
A ZigBee wireless transmitter for installation inside Robosapien.

Metrowerks CodeWarrior™ IDE) that is needed for building and testing a ZigBee design. Reasonably priced at under $200, the 13192 DSK is a great way to evaluate ZigBee with an eye on discovering Robosapien hacking applications.

Let's take a look at each of the DSK's major components in greater detail.

1. MMA6200Q Series XY-Axis 1.5g Accelerometer. Able to measure small exertion from tilt, motion, positioning, shock, or vibration forces.

2. MMA1260D Z-Axis 1.5g Accelerometer. A small package, 1200 mV/g sensitivity along the z-axis.

3. MC13192 2.4GHz RF Data Modem. This packet modem is compliant with IEEE 802.15.4.

4. MC9S08GT60 Microcontroller. A low-power, low-voltage 8-bit MCU.

What can you do with this sucker? First of all, one of the boards is set to being a transmitter, while the other one becomes the receiver (Fig. 4-18). It's a no-

FIGURE 4-18 **Two ZigBee Developer's Starter Kit transmitter and receiver boards form the backbone of a data logging sensor system that can detect tilt, axial force, and coordinate orientation, then send this data, wirelessly, to a computer.**

brainer: the one connected to your PC is the receiver, and the remote board is the transmitter. After you load some demo software, you are ready to become a data logger. And what better HOBO than your ROBO, Robosapien.

Once you've got the transmitter attached to Robosapien, your receiver/PC combo will be able to read four types of data; and, get this, these are really cool sensor readings for such a small footprint:

1. Raw Sensor Data. The on-board MCU handles all of the big number crunching, but you can view voltage output, 8-bit A/D data, and axial g force (units of gravity).

2. Angle of the Dangle. Displays the x, y, and z orientation of the transmitter.

3. Tilted. Just like that pinball game you could never beat, this reading represents the gravitational force (g) in each axis that is at an angle to the direction of rotation.

4. Fall Down. Turn the transmitter upside down, and, *bingo*, the time of this event is recorded by the receiver.

Hey, I was skeptical of this stuff at first, too. After just a little experimentation with Freescale's ZigBee DSK, however, and I could very quickly see how Robosapien could be hacked into a complete, competent home security sentry system. Yeah, no kidding. Just think, you've got this benign toy-like object that can hold a wireless video camera, accelerometer sensor, and transmitter, all three of which can be wirelessly broadcast to a remote PC. From there it's up to you. For example, have the PC phone you with a video showing an image of the intruder who just knocked over your Robosapien watchdog. Or, maybe get a farewell shot of your doorway, just before Robosapien is shoved into the trunk of the thief's car.

CHAPTER 5

I Can't Get 'em Up

I f you're like me, then you might also be a meticulous horologist. Before you shout, "Them's fightin' words," please be advised that horology is the study of measuring time, or more specifically in my case, the study of clock making.

I've always been fascinated with clock mechanisms. Pendulums, chimes, grandfathers, cuckoos, carriages, skeletons, wristwatches, analog, digital, and alarms: I've studied them, built them, and repaired them. Visitors to my house have a hard time not noticing the cacophony that fills every room every hour, on the hour. Having so many clocks does have one drawback, however—setting them. Every Sunday morning I have to set them, wind them, pull them, and crank them. It is truly a labor of love. Unfortunately, throughout the week, each clock advances or slows down to such a degree that none of the chimes, bongs, and clangs are synchronized. It can sometimes get so bad that a good 2 to 3 minutes are consumed as each clock repeatedly strikes the *same* hour. Some have referred to this phenomenon as *time standing still*. While this phenomenon can be the subject for a lively after-dinner philosophical debate, it is hardly the raison d´être for owning a clock. A clock is an instrument for measuring time and the more accurate the time, the better the clock.

A significant advancement in time measurement has been the advent of radio-controlled clocks (RCCs). These clocks are able to synchronize the exact time of day from a special radio station, WWVB, in Fort Collins, Colorado, that

THE HACK: Rewire the Robosapien power switch for making an alarm clock.

WHAT YOU WILL NEED: Battery-powered RCC ($9.99; Walgreens) (Fig. 5-1), SCR, 1N914, 4.7K resistor, 3.3K resistor, 2N2222, SPST toggle switch, relay, and one 9-volt (V) battery.

FIGURE 5-1 **A low-cost RCC for hacking into Robosapien.**

HOW MUCH: $15.35.

TIME HACK: 5 hours.

SKILL LEVEL: Butter-Bar Lieutenant.

is maintained by the Time and Frequency Division of the National Institute of Standards and Technology (NIST). Pumping out a 50-kilowatt (kW) signal on a frequency of 60 kilohertz (kHz), WWVB is able to transmit a time synchronization standard to an RCC with a stated accuracy of within 1 second or less. Even better, once an RCC is turned on, the user can just walk away and forget it. No time setting is required—ever. Now that's my kind of clock.

There are some issues with time zone settings, daylight saving time, and time code accuracy, but these problems are more properly attributed to the RCC rather than WWVB. Basically, any given RCC should have a capability for selecting a specific time zone. In the United States, this requirement would include seven time zones. On the global market, however, since an RCC isn't governed by any legal requirement, international time zone settings could be reduced or absent all together.

Additionally, a quality RCC will automatically determine the status of daylight saving time. This type of information is carried by the WWVB signal, but some low-priced RCCs will opt for the user to make this determination via a manual switch. Forgetting to flick this switch could result in a time error. If you really want to test your RCC, then check out how it handles the time during those days when we change from standard time to daylight saving time, or vice versa. This transition, as NIST calls it, should occur at 2 a.m. local time. A good RCC won't even miss a beat, but a chintzy one will probably require you to make the change.

Finally, did you know that leap seconds are periodically added to the WWVB transmission? These insertions typically occur on June 30 and/or December

Listening to WWVB

I once knew this guy in the military who bragged about listening to WWVB every morning before coming to work. I asked him why, and he replied, with a straight face, to keep himself on time. Huh? It must've been his inner cyborg self that needed the time hack. For this fellow to actually convert the WWVB pulse into something meaningful, he would've had to listen to a full 60-second (s) code frame to keep himself on time.

One of these complete time code cycles consists of an on-time marker (OTM), which is sent every second by lowering the 60-kHz frequency by 10 decibels (dB) in synchronization with a Coordinated Universal Time (UTC) second. The bit truth condition is identified by the duration of the low power OTM. For example, a 0 bit is held low for 200 milliseconds (ms), whereas a 1 bit is held low for 500 ms. Frame markers are then sent every 10 s and held for 800 ms.

If you'd like to learn more about the soothing sounds of WWVB, then point your Web browser to: http://tf.nist.gov/general/publications.htm.

Then you'll also be able to keep yourself on time.

31. This added second actually makes the final minute of the day 61 seconds long. In order for a good RCC to properly show a leap second, the number 60 should be able to be displayed in the seconds' field of the clock. A lower priced RCC will probably just hold the display of 59 seconds for 2 seconds. Now exactly what you do with extra second is up to you. This same principle holds for the display of leap years, too. In this case, however, the RCC date display must properly indicate a leap year date as February 29.

So buyer beware.

It's too bad that Robosapien (RS) doesn't have an RCC built into it, isn't it? Well, you know what? Let's add an RCC to RS and hack a cool alarm clock out of it.

The Hack

STEP 1: TWO HACKS IN ONE. Yes, that's right; you're not only going to have to open up Robosapien, you're also going to have to take your brand new radio-controlled clock apart. Rather than focus on the disassembly of my particular atomic clock model, you should concentrate on reducing your clock to a form factor that is small enough to fit comfortably within the Robosapien chest cavity. With some brands of RCCs, this could be quite an effort. Therefore, strive on finding an RCC that is small, cheap, and, if need be, ugly. Remember the clock will be housed inside RS, so looks aren't important.

STEP 2: OPEN SAYS ME. My particular RCC only required some strong fingernails to pry it apart (Fig. 5-2). Once inside it was easy to remove the needed parts (Fig. 5-3). And don't forget to remove the battery compartment, unless you plan on driving your atomic alarm clock off of the RS main circuit board.

STEP 3: PIECES AND PARTS. Essentially, you should try to remove every component from the clock. Areas where you could run into difficulty are with the antenna and the display. This step *might* require some plastic snipping or cutting (Fig. 5-4). Luckily, these messy items will be hidden away tucked inside the RS chest cavity.

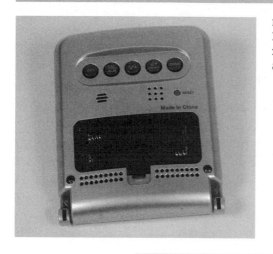

FIGURE 5-2 **Primed for prying. I had to use some strong fingernails to get this baby apart.**

FIGURE 5-3 **Inside this RCC the display was separate from the main circuit board. Note the radio antenna laying along the lower edge of the clock's plastic case. The gray plastic lid holding the RCC's buttons is a reversible lid that acts as a stand for the clock.**

FIGURE 5-4 **The alarm clock circuit is attached to the AL1 and AL2 lines, which drive the RCC speaker.**

Pulsing to an International Beat

There are other countries who employ a similar broadcast timekeeping radio station similar to WWVB. The ones that can be used with an RCC are:

China: BPC (68.5 kHz)

Germany: DCF77 (77.5 kHz)

Japan: JJY (40 kHz and 60 kHz)

Switzerland: HBG (75 kHz)

United Kingdom: MSF (60 kHz)

STEP 4: BUILD IT AND THEY WILL WAKE UP. Construction of the radio-triggered relay circuit is very simple (Figs. 5-5 and 5-6). This would be a good time for learning one of the hacker's favorite circuit board building tricks, too. You know those multipin integrated circuit (IC) sockets that you typically soldered onto a circuit board? Rather than soldering the socket to a board, use the socket *as* your board (Fig. 5-7). In this technique, you insert components between the various pins of the socket and solder them together; much like you do with a modular solderless breadboard. Try it, before you knock it.

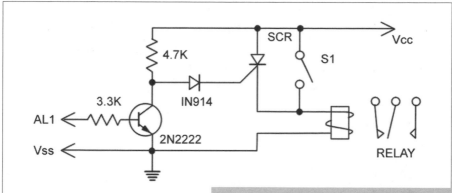

FIGURE 5-5
Schematic diagram for alarm relay circuit.

FIGURE 5-6 **Parts used for building alarm relay circuit.**

FIGURE 5-7 **Circuit on a chip.**
Integrated circuit sockets make marvelous hacker breadboards.

STEP 5: AND, FINALLY... Now wire the relay into the RS power switch located on the back chest plate (see Figs. 5-8 and 5-9). When this connection is complete, the input to the relay must be wired into the speaker output of the RCC. Before you seal the Robosapien back up, test your circuit. Set a wake-up time for about 1 minute from the current time. Now wait. Did RS go through its wake-up routine? If so, congratulations; set your normal morning get-up time, put RS back together, and get ready to wake up to a real, gruntin' bot. Ouch.

FIGURE 5-8
Carefully remove the Robosapien power switch circuit board from the back chest plate.

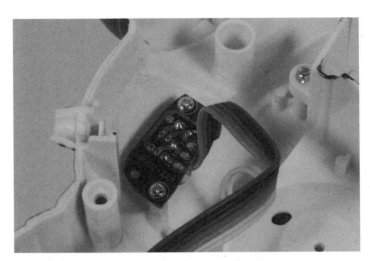

FIGURE 5-9
Connect the relay circuit from the RCC to the RS power switch.

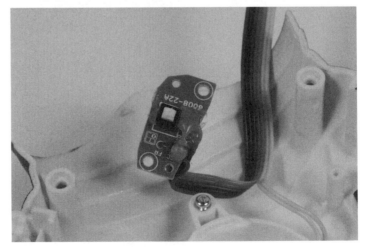

Ouch, Something's Wrong

If nothing works (neither RS nor RCC), check and recheck all of your wiring and solder connections. A multimeter is very handy for this type of diagnostic check. The biggest problem that I've had with this hack is trying to get the RCC to behave properly. Most of these troubles were traced to broken connections that I made when I opened up the clock. Therefore, it's best if you can find a clock that does *not* have to be disassembled prior to installation inside RS. In fact, if you can just snip the connections to the speaker, connect those wires to your relay circuit, and wire your hack into the Robosapien power switch, your chances for success will be extremely good.

Rosebud..., One More Thing

In contrast to the previous chapter's hacks, this advanced section is actually a dumbed-down version of the previous alarm clock hack. Realizing that the RCC and some of the associated support components could be difficult to wire into the Robosapien power switch, I present to you this low-tech option.

There is a cute little countdown/count-up timer from RadioShack stores that has some great looks and an equally great price. For less than $10, the Easy Set Timer (63-1601) is a clock with dual timers, alarm noise, and flashing LED. Furthermore, the small size of the Easy Set Timer allows it to be hidden neatly away on the back of Robosapien. While you could probably wire this timer into the RS power switch like the above hack, this alternative hack just requires you to attach the timer, set the clock, set when you want to wake up, and go to sleep. While you *won't* wake up to the sounds of your favorite robot, you will have an alarm clock that looks a lot like a Robosapien. Oh, and don't forget to remove the magnet on the back of the Easy Set Timer before you attach it to Robosapien.

CHAPTER 6

Shuffle Off
to Buffalo

Do you know the three S's of hacking? Safe, smart, and style form the credo for all hackers. In a nutshell, you should always practice safe hacking. Seriously, when dealing with power line voltages, it is better to be safe than sorry. Always test and retest your hack with a multimeter *before* you introduce live line voltages into your circuit.

Just as fundamental as being safe with your hacks, you should only spend your time creating smart hacks. For example, hacking Robosapien with a microwave oven would not only be an unsafe hack but also a stupid hack. Your design creativity is a lot better served by making smart hacks—like those included in this book. Finally, as with everything in life, style counts. No designer wants to create an ugly product. Sure, budget and time constraints can lead to cutting some corners, and the local landfill is littered with many examples. If you have any desire, however, to be a competent hacker, then remember to adhere to the three S's.

The reason for mentioning these fundamental design tenets for hacking is that many hackers fail to observe any or all of them. These are hacks, not hackers. Case in point: Right after the commercial release of Robosapien, numerous Web-based chat forums entertained a dialog about "hacking" Robosapien. Most, if not all, of these attempts were poorly prepared, weakly designed,

FIGURE 6-1 Apple iPod mini. (Photograph courtesy of Apple)

lame hardware or software construction projects that were far from hacks. In one specific instance, the winner of a widely publicized contest demonstrated everything that is wrong with bad hacking. My gosh, you don't just slap a music player on another device and claim that you're a hacker. Be smart and get some style.

One of my first hacks with Robosapien was similar to this contest entry, but instead of just slapping a junky MP3 player on poor RS, I integrated an Apple Computer iPod® mini (Fig. 6-1) with onboard stereo 1-watt (W) amplifier and dual speakers inside the robot. The result was sweet, smart, and stylish (Fig. 6-2). Like most real hackers, I wasn't satisfied, however. I didn't like losing an iPod mini to RS. So during the *Macworld* San Francisco 2005 keynote presentation when Steve Jobs announced the iPod shuffle (Fig. 6-3), I knew that

FIGURE 6-2 My first iPod/ Robosapien hack.

my Robosapien iPod mini hack was ready to be revisited and improved.

By now you've heard the size and weight of the iPod shuffle compared to the specs of a pack of gum. Well, I'm here to tell you that this music player is more akin to a car key fob and a couple of simple switches with the now familiar circular control pad than it is to a couple of sticks of gum. Even sweeter is the music that can blast out of the iPod shuffle. The $99, 512-megabyte (Mb) model with a capacity for 100+ songs and the double-your-pleasure $149, 1-gigabyte (Gb) one, which is capable of holding 200+ songs, are both equally capable of providing clear, skip-proof music for around 10+ real hours of playing. That's roughly 10 music CDs' worth of music for the entry level model.

One aspect of iPod shuffle that is great for hacking to a Robosapien is the built-in lanyard. Just a little tightening up of its length, and this lanyard can be easily looped over RS's head. You get an installation that's a snap, a terrific tune maker, and a snazzy fashion statement—all in one.

Unlike its bigger siblings, iPod shuffle lets iTunes® create its playlist, load the music, and charge its battery. All of this goes through a single USB port sans cable. Or, if you so desire, you can allocate a piece or all of the iPod shuffle to be used as a USB flash drive. Otherwise, when playing the iPod shuffle you can select between randomly playing your tunes or playing them in the order that they are stored on your player. Either way you get one great sounding, low-cost, continuously playing music system that is the ideal match for this remote, wandering Robosapien music machine hack that earns a perfect "10" for style.

FIGURE 6-3 **Apple iPod shuffle. (Photograph courtesy of Apple)**

The Hack

STEP 1: ARE YOU IN OR ARE YOU OUT? WHAT'S IT GOING TO BE?
This hack can be conducted in one of two ways: simple or fun. The simple route just requires that you wire a 3.5-millimeter (mm) stereo jack into the Robosapien speaker. The iPod shuffle (Fig. 6-4) will then play your music through the onboard speaker, in monaural sound (Fig. 6-5). Yuck. The fun hack route requires that you build a stereo amp circuit and install it, plus two 8-ohm speakers inside the RS chest cavity. Whichever hack you choose, make sure that you install your 3.5-mm stereo jack near the neckline seam between the front and back chest plates. This black-painted plastic is perfect for hiding the jack.

FIGURE 6-4 **Hot off the shelf, this is one of the first 1-Gb iPod shuffles sold and the first one ever hacked into a Robosapien.**

FIGURE 6-5 **If you use the single speaker option, you will have mono sound from the built-in Robosapien speaker.**

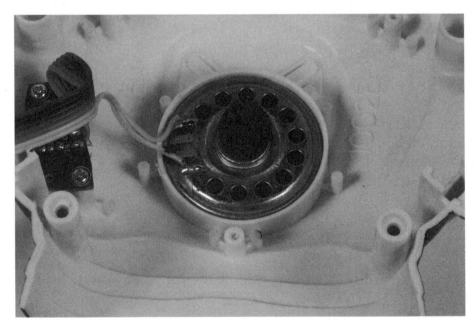

STEP 2: THIN IS IN. If you're able to locate some thin 8-ohm speakers, they should fit easily inside the front chest plate (Fig. 6-6). Alternatively, you could mount one speaker in the front plate and the other speaker inside the back plate. This would give you better stereo sound, but the back plate is a tight fit, and the speaker would have to be insulated quite a bit around the main circuit board.

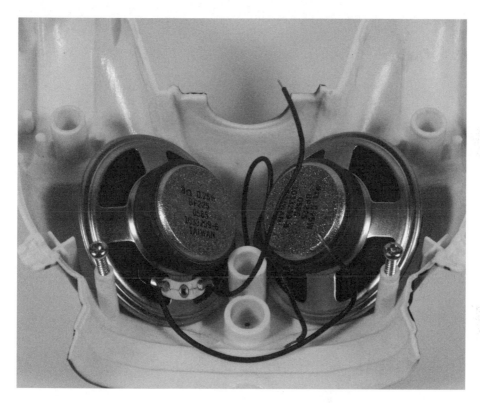

FIGURE 6-6 **Twin speakers for sweet booming 2-W stereo sound.**

STEP 3: SPEAK(ER) UP. Wire your stereo amp circuit (Figs. 6-7 and 6-8). Connect the twin speakers to this circuit, connect the 3.5-mm jack and battery, and plug in your iPod. Now adjust the 10K potentiometers for the best sound. If everything is a go, it's time for the show. Put RS back together, attach the iPod shuffle, and program RS to follow you around the house while mixing your favorite tunes. Now that's how you hack a 'sapien (Fig. 6-9)!

FIGURE 6-7
Schematic diagram for the LM386 stereo amp circuit.

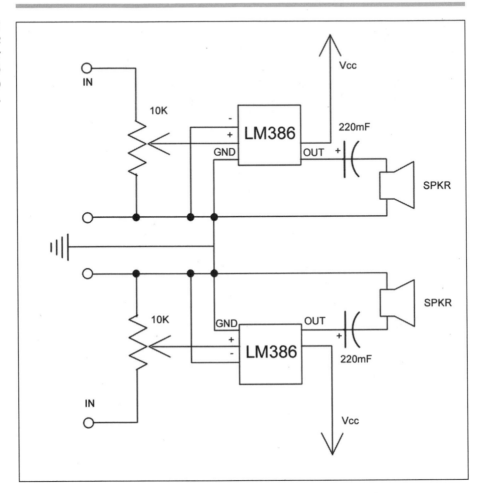

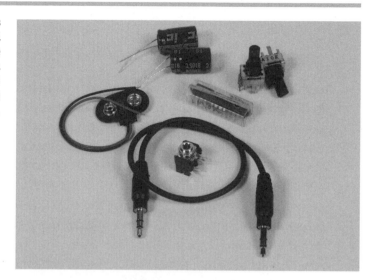

FIGURE 6-8
**Parts needed
for the
construction
of the LM386
stereo amp
circuit.**

FIGURE 6-9
**Now that's
how you hack
a 'sapien into
a sweet stereo
soundin'
'sapien.**

Mini Pod with Maxi Sound

As unbelievable as it sounds, a computer manufacturer has actually designed a product that can actually minimize a lot of the boredom associated with life. The Apple iPod mini is a svelte $3.6 \times 2.0 \times 0.5$ inches, 3.6-ounce anodized aluminum digital music player with a 4-Gb or 6-Gb hard disk drive, which Apple claims can hold 1,000 and 1,500 songs, respectively. Additionally, the iPod mini comes with a dual-function FireWire (and USB 2.0) cable that can both charge the iPod's battery and shuttle music back and forth from the included iTunes Music Store software. The iTunes software acts as a digital music jukebox that is capable of buying digital music from the Apple Music Store, converting CDs into iPod-aware AAC-encoded music, and managing your digital music database (e.g., copying music to the iPod mini, creating playlists, deleting music from the iPod mini, etc.).

A patent-pending, touch-sensitive, solid-state Click Wheel is used for scrolling through all of the iPod mini's functions and for selecting and playing songs from your loaded music collection. Rounding out the iPod mini is a reported 8-hour battery life, 1.67-inch monochrome LCD with LED backlight, earbud headphones, ac adapter, and belt clip. Finally, you can make your own fashion statement by selecting the best iPod for your hacking persona—silver, gold, pink, blue, and green are available. In initial shipments of the iPod mini, the blue anodized aluminum model was the leader in advance sales.

Even though Apple based its storage capacity claim on a 4-minute-per-song file size, my testing of the iPod mini 4-Gb version, for example, resulted in a random, "real-world" music sampling being able to store 946 songs on the 4-Gb hard disk drive. In fact, if your music library contains a large number of longer-length recordings (e.g., classical music), you should expect your iPod mini to hold even fewer songs.

Likewise, my playback tests could only squeeze 4^1/$_2$ to 5 hours' worth of music out of a fully charged iPod mini. Although the battery strength indicator did display a reduced charge, no other type of warning message was displayed prior to the iPod mini shutting down from lack of power. (*Note*: In fact, although the battery strength icon was showing a completely exhausted battery, I was still able to listen to Green Day's *American Idiot* in its entirety before my iPod went dead—that's more than 45 minutes worth of music. A subsequent attempt at starting up the iPod mini, however, did display a brief message stating that it was out of power.) Recharging the completely drained iPod mini battery took 2 hours and 15 minutes.

Unlike these "technical specification" tests, the two areas where the iPod mini really earns its Must-Buy status is with the included belt clip and iTunes software. Whether you are attaching your iPod mini to your belt or hanging it from a shirt pocket, the beautifully engineered belt clip will guarantee that your diminutive digital music player will keep on channeling great music to your ears no matter where you do your hacking.

And using the iTunes software is as effortless as riding a bicycle. Using a simple point, click, drag, and drop interface, you can build a custom playlist featuring your favorite tunes in just a matter of minutes. Then just drag this playlist into your iPod mini, hop on your bike, make a couple of selections, and you're off to the races. It couldn't be much simpler and still have its same kind of enormous music management capability. Best of all, iTunes will work on either a Mac or a PC.

Ouch, Something's Wrong

Actually, this hack is fairly easy to diagnose; if you've got a problem, it's almost always the stereo amp circuit. Check your wiring and solder connections with a trusty multimeter. Once it all checks out, power it up and try again. Just avoid increasing the volume too high, or the speakers might jump out of place and short the main circuit board. Ouch.

Rosebud..., One More Thing

Unlike the previous hacks in this book, there is very little that can be improved on the iPod shuffle hack. You could add an MP3 player circuit based on the STMicroelectronics STA013 MP3 Decoder Chip with an external compact flash card. But, you know what? That wouldn't be very smart, because the iPod shuffle embodies all of these features in a tidy, stylish form.

CHAPTER 7

Tanks for the Memories

Somewhere there is a building in a war zone that could be harboring enemy combatants. To survey this urban location, an advanced reconnaissance team is sent into the structure. The recce team encounters a steep staircase and slowly begins to climb each stair. Suddenly, hostile fire erupts into a deadly ambush, and every member of the reconnaissance team is killed. And that's just the way the U.S. Army wants it. What? You see, the recce team comprises a group of advanced tactical mobile urban robots, or "Urbies." Using some advanced vision capabilities the Urbie was able to pinpoint the enemy without endangering any human lives. Assessing the resulting data, field commanders can then make an informed decision on dealing with the enemy threat.

Armed with real-time stereoscopic "machine vision," the Urbie is a rugged mobile reconnaissance platform whose "eyeballs" are being developed by the Machine Vision Group of the Jet Propulsion Laboratory (JPL) at the behest of the Defense Advanced Research Projects Association (http://robotics.jpl.nasa .gov/tasks/tmr). Assisting with this development are IS Robotics, the Robotics Institute of Carnegie Mellon University, Oak Ridge National Laboratory, and the University of Southern California Robotics Research Laboratory.

Unlike previous guided robots, the Urbie is an autonomous, free-roaming robot that is able to interact with its stereoscopic vision. This interactive capability enables the robot to quickly and accurately navigate toward a programmed target while seeing and avoiding obstacles along its way. Therefore, human operators don't need to watch, interpret, and control the robot's actions.

This autonomy enables the Urbie to be adapted for a wide number of different roles. From wartime conflicts to urban disasters, an Urbie could be used for entering and surveying structures, buildings, waste sites, urban wreckage, and confined spaces where human observation would be dangerous, if not impossible.

Another more mundane capability of the Urbie is its unique tractor system. Using a quad tread design, the Urbie is equally able to scamper up a staircase; charge down a drainage ditch; and sprint through mud, water, dirt, and dust (Fig. 7-1). This is the perfect mobility system for an autonomous robot that could encounter just about any kind of obstacle in a hostile environment. And let's face it: Most hostile environments aren't carpeted. The tracked platform of the chassis is called "Urban II." This portion of the Urbie was developed by IS Robotics (now iRobot®). In addition to its more conventional tracked side treads, the Urbie also has dual tracked articulations along the front outside of each main tread. Know as QuickFlip™, iRobot likens these articulations to flip-

FIGURE 7-1
Urbie— the tactical mobile robot. (Photograph courtesy of Jet Propulsion Laboratory)

pers. These articulations can do continuous 360-degree rotation, helping the Urbie climb stairs, scale curbs, and surmount rocks.

The treads, themselves, are made from a tough flexible polymer called ToughTrac™. A special design enables the treads to eject debris from underneath each tracked element, while making a sure-footed contact with any type of surface, even those tough "slippery when wet" surfaces. While ToughTrac and QuickFlip mobility systems are too pricy for hacking, this same concept can be applied to a new pair of "shoes" for Robosapien. Be forewarned, this new set of tracked feet will make your RS considerably faster than its current waddling self. So let's get to the races.

The Hack

STEP 1: GET TANKED. Up front, I would like to tell you that these armor kits from Academy represent one of the best bargains in robot building. Two motors complete with gearboxes are mounted in a neat, self-contained chassis. Even if you don't intend to build a tracked robot, this motor/gearbox combination can be used in a variety of other projects (Fig. 7-3). Finally, this manufacturer's tank models do not require any cement for assembling. This important construction element is extremely important when you need to add, remove, or change batteries inside each tank chassis (Fig. 7-4).

Before you begin any major wiring, you should consider painting your future tank tread feet. As shown in Fig. 7-5, I opted for glossy white with black treads and bogie wheels. Your paint should dry completely before assembling the tanks. Assemble each tank as described in the instructions, except for two wiring modifications. First, don't attach the tank's remote control to the motor/gearbox. Instead,

FIGURE 7-2 **These two model AFV kits will supply a new means of traction for Robosapien.**

you will be soldering four pieces of 18-inch-long wire to each motor terminal. Second, route these four wires up through each tank's turret pivot. You might have to drill a hole in the tank chassis for completing this modification.

FIGURE 7-3 **This is one of the best motor/gearbox buys on the market. And the manufacturer throws in a fun tank kit, too.**

FIGURE 7-4 **You don't need any cement for assembling these tank kits. Therefore, you get easy access to the motor/gearbox, batteries, and wiring after this hack is completed.**

FIGURE 7-5
**A glossy white
paint finish
makes the
final hack
look better.**

STEP 2: MY LEGS—I CAN'T FEEL MY LEGS. Remove both Robosapien feet (both inner and outer plastic plates) and both battery compartments (Fig. 7-6). The power for both, the tanks and Robosapien, will be derived from the batteries used in each tank chassis. Unlike the walking Robosapien, this tank tread hack does *not* require any rocking action at the knee joints. Therefore, you will need to "freeze" this joint. I was able to stiffen this joint by gluing the Robosapien plastic in two places; first glue the hip joint to the main body thigh, and second, glue the knee joint to the top of the tank chassis. I used a thick cyanoacrylate glue for this step. Once the glue dries, you should have a rock solid leg/foot assembly.

FIGURE 7-6
**I ain't goin'
anywhere now.
Robosapien
is ready to
be adapted
to a new set
of dogs.**

TANKS FOR THE MEMORIES   **109**

Tanks U

Armor kits, armored fighting vehicle (AFV) kits, or tank kits come in many different sizes (called scales), nationalities, and manufacturers. When hunting for a kit that can be used for providing a tracked chassis, keep in mind that most model tank kits are not motorized. This type of kit is for scale model builders who strive for accuracy over mobility. In my hacking of various motorized AFV model kits into viable robots (with varying degrees of success, mind you), I have come up with this short list of manufacturers that produce motorized tank kits:

Academy Hobby (www.academyhobby.com): Korean kits; best for general robot hacking; 1/48 scale remote control mini tank series.

> ➤ Merkava MBT
> ➤ T-72 Russian Army MBT
> ➤ Challenger British MBT
> ➤ Leopard 2 A5 MBT
> ➤ M1A2 Abrams MBT
> ➤ Leclerc French Army MBT

Micro-X-Tech (www.dragonmodelsusa.com): Hong Kong kits; tiny, detailed kits requiring advanced hacking skills; 1/72 scale radio-controlled series (these kits can be difficult to find; you could consider the Hobbico 1/35 scale radio-controlled combat armor tank set as an alternative; www.towerhobbies.com).

> ➤ M1A1 Abrams (different radio frequencies)

Tamiya (www.tamiyausa.com): Japanese kits; outstanding quality for high-performance robot hacking; 1/35 scale radio-controlled series (1/16 scale series of AFV model kits are much bigger and much more expensive).

> ➤ Leopard 2 A5 Main Battle Tank
> ➤ King Tiger

STEP 3: HMM, NOW WHICH WIRE? OK, this step is much easier than it might seem. All you're going to do is wire the Robosapien power grid to each tank's battery compartment and reroute power from the Robosapien hip motors to each tank's motor/gearbox. Got it? Well, for the wiring-challenged hacker, the left Robosapien battery wiring is a dual violet/black wiring harness, and the right Robosapien battery wiring is violet/orange in color. Additionally, the left hip motor wiring is a twisted orange-blue pair, while the right hip motor's wires are a twisted orange/tan pair. You can just disconnect the Leg-L and Leg-R sockets for rerouting the motors. However, you will have to do some snipping of the battery wiring. Just get as tight as possible to the original Robosapien battery compartment wiring, close one eye, and snip. Check all of your wiring, snipping, and soldering before you seal up the twin tanks and your RS.

STEP 4: BATTERY CRAMPS. If the stock battery pack doesn't fit inside the tank chassis, you will have to install a new battery power system. The 3-volt (V) lithium batteries found in some digital cameras, for example, are an ideal substitute for the model kit's battery pack (Fig. 7-7). Install one lithium battery inside each tank chassis.

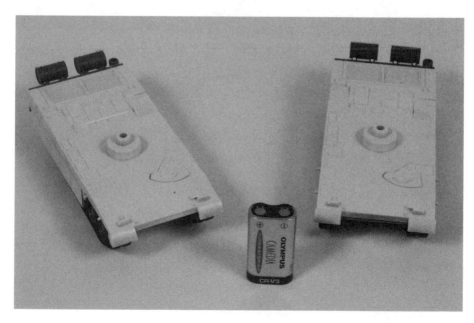

FIGURE 7-7
A 3-V lithium battery pack is perfectly sized for fitting inside each tank chassis.

STEP 5: ROLLIN', ROLLIN'. Feeling cocky, are ya, punk? Slap your batteries into each tank chassis, switch on the Robosapien power, grab the infrared (IR) remote control, and issue a leg movement command. Everything OK? Good; now get out there and make some tracks (Fig. 7-8).

STEP 6: WHAT'S THE FREQUENCY, KENNETH? The performance of the tank motors can be enhanced by altering the Robosapien main circuit board crystal capacitor. Refer to Chapter 11 for more information about this hack.

FIGURE 7-8
These tanks are made for rollin' and that's just what they're goin' to do.

Tarantula

Imagine my surprise when I was nearing the completion of my tank track tread feet hack and I saw the MGA Entertainment R/C Tarantula (Fig. 7-9). There on the local toy store shelf was a remarkable copy of an Urbie—for the rest of us.

FIGURE 7-9 The MGA Entertainment R/C Tarantula is equipped just like the Urbie with four articulated moveable treads, but it costs thousands less. The Tarantula retails for less than $100.

Just like its Department of Defense (DoD) counterpart, the Tarantula utilizes articulated, moveable tread feet for navigating along the most difficult terrain. I doubt that its plastic body is as bulletproof as the Urbie's, however. Completely radio controlled with six directional functions (e.g., forward, backward, left turn; both forward and backward, and right turn; both forward and backward), as well as working headlights, the Tarantula can climb a set of dark stairs with the ease of its working-class sibling. The best part about the Tarantula, though, is that it costs thousands of dollars less than an Urbie. Hey, Uncle Sam, buy yourself some Tarantulas so that you'll have enough money left over to buy a couple of those $150 DoD hammers.

STEP 7: LET'S GO CRAZY HERE. Once this alternate mobility bug bites, it bites hard. I found two excellent replacements for Robosapien's feet that surpass even the tank treads. First, MGA Entertainment has an RC vehicle that can both drive on land and propel across water (Fig. 7-10). As shown in Fig. 7-11,

FIGURE 7-10 **Land Sea 2 from MGA Entertainment is an RC vehicle that can both drive on land and sail on water.**

FIGURE 7-11 **Hey come back here. This is the new all types of terrain robotic vehicle (ATTRV), but you won't find it in any store—so hack one.**

the MGAE Land Sea 2 is a great match for the dimensions of Robosapien. Likewise, MGAE's Shadow Shifter (Fig. 7-12) is not only a great looking vessel in its own right, but as Fig. 7-13 shows, it's also a sleek looking pair of powered skis for Robosapien. Let's go surfin' now.

FIGURE 7-12
MGA Entertainment's Shadow Shifter featuring the Sea Prowler 99™ is cool even without adding Robosapien to it.

FIGURE 7-13 **Now where's Gidget? Watch your transmitter's range with this baby or you'll be swimming out to bring Robosapien back to shore.**

Urbie, LEMUR, and Sojourner, Too

There is quite a rogues' gallery of robots that have been developed at the JPL. To any robot enthusiast, however, these names read like a who's who list of famous robots. In no particular order, this distinguished membership includes **LEMUR, Spiderbot, Nanorovers, Hopping Robot, Cryobot, FIDO, Sojourner, Snaking Robot, Hydrobot, Urbie, Rocky 7, and SRR.** You can read all about these celebs at http://technology.jpl.nasa.gov/gallery/index .cfm?page=imagesAllByCat&catId=10.

Ouch, Something's Wrong

Good golly, a lot can go wrong with this hack. For starters, snipping the wrong wire is a bad thing. Also, soldering the wrong RS motor wire to the wrong tank motor/gearbox terminal will result in the tank(s) running backward. This polarity error can be downright bewildering to troubleshoot when only one of the tanks is wired incorrectly. If you have a polarity problem, you can try this old hackers' trick to help finger the culprit. Determine which tank you think is running backward. Now remove its batteries and reinstall them backward. If the tank now behaves properly, you can either make a note to always insert the batteries backward, *or* resolder your connection with the correct polarity. Me, I usually just put the batteries in backward and call it a design feature.

Rosebud..., One More Thing

After you've driven your track-equipped RS around for a while, you should notice two things: There is a significant speed increase over its former walking self, and Robosapien now has a tendency to run into obstacles. Although the former problem will be a factor of the type of tracked system that you use (i.e., some tread systems could be faster than others), the latter problem is due to the loss of the RS toe and heel touch sensors. With just a little effort, you can remove the touch sensors from the discarded Robosapien feet and integrate them into your new tracked mobility system.

Remember folks, let's practice some style here. So don't just slap the toe and heel touch sensors onto each tank chassis. Rather, modify the installation with some parts that are left over from the tank kit's turret assembly. A little thought at this point can go a long ways toward making a hack that you will be proud of instead of another resident of your wastebasket.

CHAPTER 8

Get This Monkey Off My Back

Are you worried about our continued dependency on fossil fuels? If you are, then chances are that you have considered buying, if you have not already purchased, a hybrid automobile. If you're new to this notion of reducing your consumption of gasoline, then think of a hybrid automobile as basically a regular vehicle with two (or, more) forms of power. Generally speaking, the two power sources are a gas combustion engine and a rechargeable electric battery/motor. Although there are other forms of hybrid power, including bio-diesel and hydrogen, the most common form used by Ford, Honda, and Toyota is the gasoline engine/electric motor hybrid.

During normal daily driving, a hybrid automobile's gasoline engine shuts off in low-speed situations, typically less than 25 miles per hour (mph), as well as when stopped at traffic lights. In these situations, the power for the vehicle is supplied by the electric motor, except when stopped, then both power sources are off. To recharge the electric battery, hybrid automobiles will recover the energy generated during braking for charging. This form of operation is sometimes called regenerative braking.

More accurately, when a vehicle can selectively operate on its gas combustion engine, or its electric motor, or a combination of both power sources, this type of hybrid is called a full hybrid. This form of driving enables a hybrid to provide good fuel economy, acceptable performance in city traffic conditions, and, most important, reduced exhaust emission. Therefore, a hybrid is good for both ends of the environment spectrum—minimal impact on fossil fuel supplies and moderated contribution to air pollution.

One of the real improvements with hybrid designs has been the introduction of regenerative braking. Regenerative braking works by reversing the direction of the electric motor, so rather than using electricity to turn the wheels, the rotating wheels turn the motor and, therefore, create electricity. Another by-product from regeneration braking is that using energy from the wheels to turn the motor, in turn, slows the vehicle down.

Without a doubt, though, the feature of a full hybrid that is completely unnerving to every seasoned gas combustion engine driver is that when the vehicle comes to a stop, as at a red light, both the gas combustion engine and the electric motor shut off automatically, so that energy is not wasted idling. My gosh, this thing just died. Well, not really, the battery continues to power auxiliary systems, however, such as the air conditioning and dashboard displays. Thank goodness.

Wow, wouldn't this be a great capability to incorporate into a robot project? Several years ago I designed a circuit built from a surplus electric motor, which, when the motor was manually spun, would charge a capacitor that, in turn, would recharge a NiCad. Even with a pretty good amount of gear reduction, this process took way too much cranking to be practical. The notion, however, was to be able to recharge a NiCad in a remote location. By replacing the motor portion of the circuit with a bank of photovoltaic cells, the result was the same, but the process was much smoother and efficient.

Now imagine if you wanted to create a hybrid Robosapien—what would you do? How about hack another robot's attributes into RS? In this case, OWIKITS WAO Kranius is the donor, and Robosapien the recipient. Hey, this hack really rocks.

The Hack

STEP 1: BRAINS OVER BRAWN. Unlike the previous hacks, the final result from this hack will bear absolutely no resemblance to Robosapien. Begin by removing the main circuit board from your Robosapien (Fig. 8-2). Set this board aside, it's your future hybrid robot's brain.

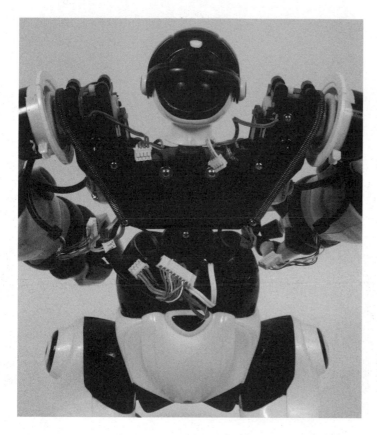

FIGURE 8-2
Brainless and dangerous. The main circuit board has been removed from Robosapien for subsequent installation in WAO Kranius.

STEP 2: GET YOUR HANDS DIRTY IN ROBOT INNARDS. Begin the construction of your WAO Kranius robot (or, substitute robot kit) (Fig. 8-3). Assemble the main body; build the gearbox; add the pulse sensor, floor sensor, and speaker; and install the drive system (Fig. 8-4). Do not, however, add either the keyboard or the WAO Kranius main PC circuit board. You will also need to retain the battery box for installation later.

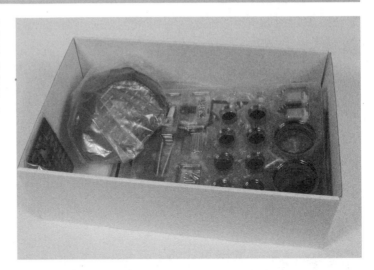

FIGURE 8-3
You'll only need a small No. 0 Phillips screwdriver for building WAO Kranius.

FIGURE 8-4
The pared-down WAO Kranius is ready for receiving the Robosapien main circuit board.

STEP 3: ABBY NORMAL. With Robosapien brain in hand, connect the WAO Kranius sensors, speaker, gearbox, and power supply to the equivalent plugs on the RS main circuit board. This process is easier than you might think.

STEP 4: HEY, HOW DO YOU CONTROL THIS THING? Remove the infrared (IR) receiver sensor from the helmet of Robosapien. Likewise, retain the attached six-LED circuit board from inside the helmet. If you wish to control your hybrid via the Robosapien IR remote control, both of these circuits must be installed in WAO Kranius. Any one of the keyboard openings on the WAO Kranius rear panel are ideal for holding the IR receiver. This arrangement will give a clear, unobstructed reception for the IR signal.

STEP 5: A JOLT OF VOLTS. Just double-check all of your wiring and connections, then insert some batteries into your newly created RoboKranius. The original WAO Kranius is a battery hog. While the 9-volt (V) battery is perfect for WAO Kranius, it's overkill for RoboKranius. Furthermore, since we aren't powering the Robosapien skeletal and musculature systems, 3 V should be adequate for powering RoboKranius. Insert two AA-size batteries and power up your hybrid (Fig. 8-5). Now see if the IR remote control can issue a recognizable command to RoboKranius. Watch out, however, or this newly planted hybrid "bug" could generate a Picosapien, a RoboBotster, or a RoboDog.

FIGURE 8-5
OK, it doesn't walk, but it sure does have a funny movement pattern; kind of like a late, late Friday night walk home.

Ouch, Something's Wrong

Gee, what couldn't go wrong with a fusing of two entirely different technologies together? I had a hard time getting the WAO Kranius sensors to work with the RS main circuit board. If you find yourself having similar troubles, just disconnect the sensors and try using the RS IR remote control again.

Also, remember, WAO Kranius doesn't have any arms. Therefore, the RS arm and hand commands won't be operable. Try to focus on sounds and forward and backward movements. Slowly expand your experimentation into simple turns. Who knows, maybe you can get the sensors to work. I had pretty good success by keeping the RS microphone attached to the main circuit board and programming the sound sensor for reverse control. This hybrid hack is a great model for testing many robot building techniques. Just don't give up!

Rosebud..., One More Thing

Although you might not have a WAO Kranius for this hack, you could probably think of a good substitute for your own hybrid. In no particular order (these are all outstanding examples of robot DIY kits that are perfect for hacking), here is a short list of prime candidates for a Robosapien hack hybrid:

iBotz Picobotz (Fig. 8-6)

JoinMax Digital Tech IQ Series: Smart Arm, Quadriped, Hexapod Monster, and Robot Dog (this is my personal favorite—the hacking possibilities of this series are awesome) (Fig. 8-7)

Parallax Boe-Bot (Fig. 8-8)

Robotic Connections Botster with EmbeddedBlue (Figs. 8-9, 8-10, and 8-11)

Hybrid Horseflesh

Currently, there are five commercially available popular hybrid models from three different manufacturers. As gasoline prices soar, look for more models and some intriguing concept cars to be released in the future. Here's a quick hybrid model scorecard:

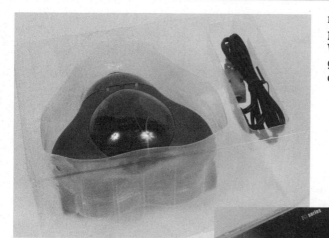

FIGURE 8-6 **Picobotz is a perfect replacement for WAO Kranius. Plus you get to keep the cool blue color in the final hybrid.**

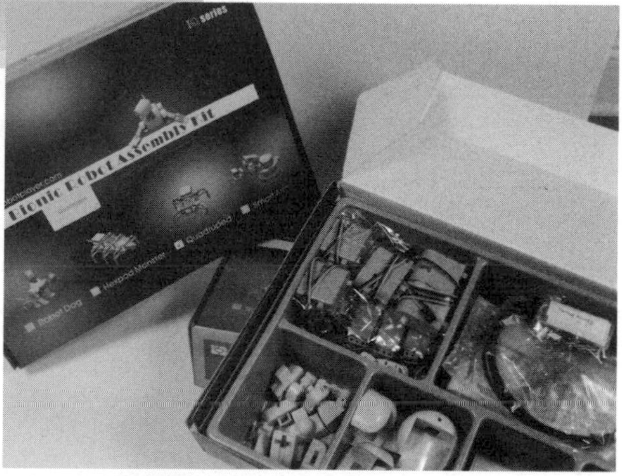

FIGURE 8-7 **JoinMax Digital Tech kits are outstanding resources for the hacker. Pictured is the open box of the fabulous Robot Dog kit. In the upper right corner of this box is a compartment filled with servo motors. Just how many motors? Try 15.**

FIGURE 8-8 **The venerable Boe-Bot is a good starting point for seeing what makes the Robosapien main circuit board tick.**

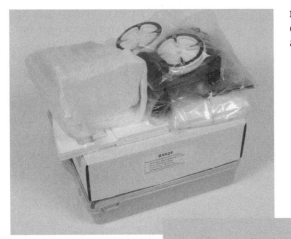

FIGURE 8-9 **Botster consists of a compact set of easily assembled pieces.**

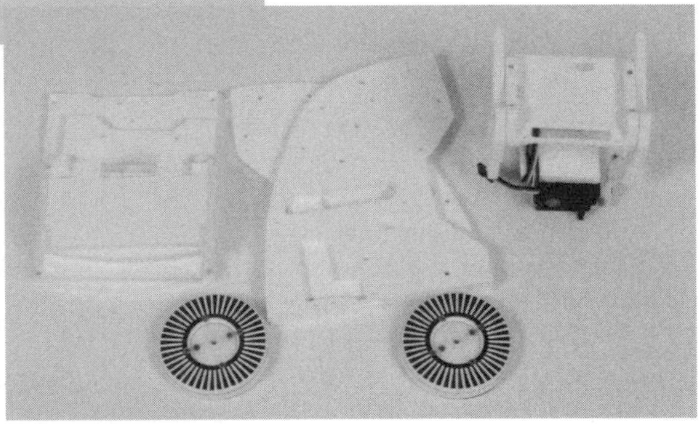

FIGURE 8-10 **Most of Botster is formed from precision-milled high-grade plastic.**

FIGURE 8-11 **One of the newest options for Botster is the EmbeddedBlue add-on, which provides Bluetooth wireless technology.**

Ford Escape Hybrid: FWD, 4 × 2, 36 city/31 hwy; and 4WD, 4 × 4, 33 city/29 hwy.

Honda Accord Hybrid: 29 city/37 hwy.

Honda Civic Hybrid: CVT, 48 city/47 hwy; and MT, 46 city/51 hwy.

Honda Insight: CVT, 57 city/56 hwy; and MT, 61 city/66 hwy.

Toyota Prius: 60 city/51 hwy.

Design It and They Will Roll

Did you know that one of the new Erector sets from Meccano S.N. features a radio-controlled chassis along with flexible spring steel parts? Complete with a radio, rechargeable battery, and more than 400 parts, the $100 RC 4 × 4 Truck set (#848700) is a remarkable departure from the more conventional Erector sets (Fig. 8-12). Granted, many of today's Erector sets come equipped with either a powerful 3-V or 6-V electric motor, but this set is the first Erector toy to be radio controlled.

FIGURE 8-12
The newest Erector set is the RC 4 × 4 Truck set.

You can build three different off-road type vehicles with this set. Each model is attached to a preassembled chassis. This chassis contains the motor, gear box, and radio receiver. All four wheels and the two axles are also attached to the chassis, so you can literally run your design right out of the box. As a robot builder, however, I really loved the new flexible spring steel parts included in this set. Now I can painlessly make some great radius-type robot designs from these pieces combining them with some of a conventional Erector set's metal parts. Along with the famous Erector set machine bolt and nut joining system, the RC 4 × 4 Truck set is a great starting point for hacking a four-wheel drive hybrid Robosapien design as shown in Fig. 8-13.

FIGURE 8-13
You can either control the truck with he RC unit or wire it into Robosapien as described with the twin tank tread chassis hack in Chapter 7.

That "Other" Amorphous, Blue, Blob, Bot

Other than OWIKITS WAO Kranius, there is another blue robot that could be a refuge from the 1958 movie *The Blob*. The sub-$3,000 Roborior, spawned from a collaboration between Sanyo and tmsuk, isn't your typical educational robot, however (Fig. 8-14). No; Roborior is your new digital watchdog. Built inside Roborior is a camera, wheels for moving around your house, a series of color changing lights (for indicating Roborior's mood, no less), and a special cell phone remote control capability. Actually, it is these last two features that drive Roborior's home defense system.

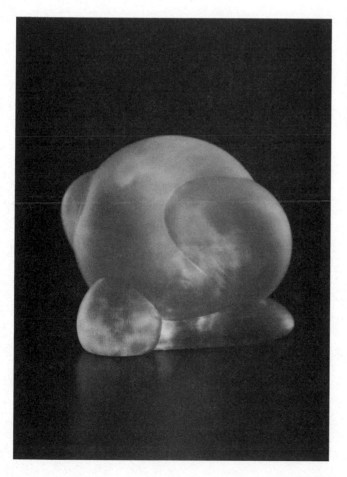

FIGURE 8-14
Ah, a peaceful Roborior roaming your home keeping the bad guys at bay. (Photograph courtesy of tmsukSANYO Co., Ltd.)

Say, for example, a Steve McQueen impersonator breaks into your home while you're away watching a late-night *Star Wars* movie festival at the local cinema. Well, Roborior will detect the intruder, snap a picture, dial you up on your cell phone, and send you the photograph of the burglar. You call the cops and keep on munching on your popcorn. Gee, thanks, Roborior. Oh, and did I mention that Roborior will glow red when it detects the intruder? Sure beats barking, doesn't it? Finally, in case you're wondering, in a nod to the robot's designer (Me Company) the name *Roborior* is derived from the melding of robot and interior. Get it?

CHAPTER 9

Vac Me Away

Wow, it has got to be a robot builder's dream come true: A domestic robot that tirelessly cleans your home. In this case, the robot is the floor vacuum system, Roomba from iRobot. Unlike previous attempts at household robots, Roomba has been a financial success for iRobot. In October 2004, iRobot reported that 1 million Roomba Robotic Floorvac systems had been sold. Not to be confused with the *Saturday Night Live* advertisement parody, Woomba, Roomba is actually two different systems interfaced together in a neat consumer package.

First, and foremost, Roomba is a robot. Housed inside a low-profile, durable plastic shell, Roomba combines an effective sensor system with a microcontroller brain that is able to provide, what the folks at iRobot call, cleaning intelligence. Regardless of your love for robotics, cleaning is really the name of the game for a floor vacuum system. Right?

A revolutionary cleaning system is the other Roomba success story. Although a typical household upright or canister vacuum system is effective for removing heavy dirt and fiber materials, it is overkill, like making omelets with a sledgehammer. Sure it breaks the eggs, but at a huge cost. Plus these conventional vacuum systems are limited in their abilities to clean under low-height areas around furniture and under beds. Roomba, however, is a completely mobile, low-profile cleaning system that can crawl under your bed as easily as the dust bunnies hiding next to your socks. Furthermore, it operates continuously so there is no user fatigue factor that sometimes results in sloppy cleaning patterns. Roomba cleans your entire room without stopping for furniture moving or varying height adjustments for different types of floor coverings (e.g., tile, wood floors, carpet, etc.).

So what if Roomba gets stuck? There is a special "stasis," or stuck, program feature that is executed when Roomba finds itself between a rock and a hard spot. It's like a squirming reverse movement program that travels backwards along the same path that got Roomba in trouble in the first place. Additionally, there is a stair avoidance feature and the ability to follow walls. This latter feature is pretty good for sucking up dirt that typically blows up against a baseboard.

However, if you want to keep your Roomba under control and away from dangerous obstacles, then you can simply place a "Virtual Wall" in its path. A Virtual Wall is nothing more than an oversized infrared (IR) beam that signals Roomba to back up and stay clear. Likewise, the Virtual Wall is a big battery waster. Two D cells are used to spray an IR beam at three different power settings across an opening (Fig. 9-1). Rather than wasting batteries on a Virtual Wall, let's waste them on another Robosapien hack. This time around, Robosapien will be our Virtual Wall.

The Hack

Rather than list a bunch of theoretical steps, let me level with you, right up front. I couldn't get this hack to work. The sample that I received from iRobot was filthy and unworkable. Therefore, I had to settle on speculating that this hack should work. Basically, your goal should be to transform Robosapien into a Roomba Virtual Wall. To achieve this goal, you will need to disassemble the Virtual Wall and install its IR components inside RS (Fig. 9-2). Once you've completed the hack, station RS on sentry duty outside an area that you don't want Roomba to clean. Now sit back and watch the sparks fly when the housecleaning robot meets the boss of the house.

FIGURE 9-1
Hey, you'd better stop right there, buddy. A Virtual Wall keeps a Roomba Discovery out of an area that shouldn't be vacuumed. (*Note*: The shading is a pictorial representation of the IR beam emitted by the Virtual Wall.) (Photograph courtesy of iRobot)

FIGURE 9-2
You can corral your Roomba Discovery into a user-defined area with the IR beam-emitting Virtual Wall. (Photograph courtesy of iRobot)

Ouch, Something's Wrong

From my cursory study of this hack, the biggest chore will be reducing the bulky Virtual Wall into a small enough form factor so that it can be inserted into Robosapien. Alternatively, you could experiment with the Virtual Wall's IR signal (using our trusty friends IRTrans and iRed from Chapter 2) and design a new circuit that mimics the Roomba "stay out" signal. If push comes to shove, however, you could just position Robosapien in the doorway and have a karate chop delivered whenever one of the two RS foot sensors is tripped (Fig. 9-3). Hi-yah, Roomba.

Say What?

In a language of beeps, pops, and tweets, Roomba speaks in terms that only R2D2 could love, and understand. This dazzling vocabulary includes:

Done

Stuck

Back Up

Continue

Wake Up

Finish Spot

Still Done

Start Spot

Start Clean

Stuck Code

Battery Error

Found Charger

Can't Comply

Battery Discharged

Remote Error

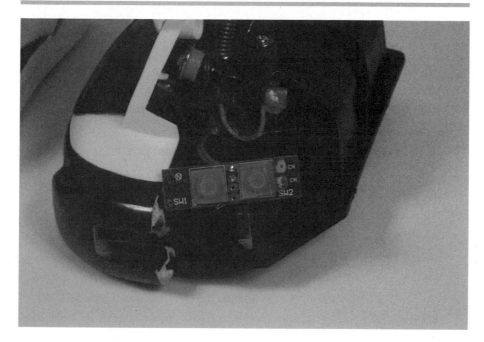

FIGURE 9-3
**One of
Robosapien's
foot touch
sensors.**

Rosebud..., One More Thing

Granted, there's not much else you can do with a floor vacuum system. Well, I suppose that you could try some sort of other goofy home appliance amalgam, but that just doesn't sit very well with me. Alternatively, I played with the notion of integrating RS with one of the popular lawn mower robots, but the results just weren't safe enough to even mention in this book. A more logical hack would be expanding the Virtual Wall concept into some sort of mobile system where Robosapien would move the IR beam to different room locations based on some sort of user implemented criteria. One of the easiest means for "telling" RS to move would be with the alarm clock feature discussed in Chapter 5. Then with Roomba and Robosapien moving in concert to clean your home you would have more time to, er, mow the lawn.

Stop Fingering Me

While fiddlin' around with this home appliance hack concept, I stumbled upon a great, low-cost touch sensor that can be integrated directly inside Robosapien (or, any other robot project, for that matter). The QTouch™ family of touch and proximity sensors from Quantum Research Group have the ability to sense physical contact through plastic or glass up to 100 millimeters (4 inches) thick. Even gloves won't stop QT110, QT111, QT112, QT113, QT115, and QT118H from detecting a touch. Even better QT113 and QT118H are also be able to sense moisture. So sweaty hands, beware.

The QT11x family of chips are as easy to use as they are inexpensive. Just slap an external sampling capacitor on it, add a detecting electrode, and zap it with around +3 volts, and you have the perfect robot touch sensor (Fig. 9-4). Furthermore, the chips all feature autocalibration, drift compensation, and automatically calibrate themselves after a time out.

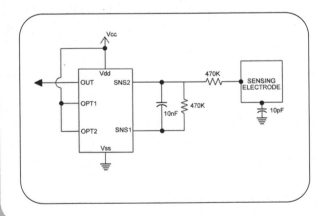

FIGURE 9-4
Schematic diagram for a typical QT11x touch sensor circuit.

As for applications, QT11x chips are found in a wide variety of commercial, real-world devices from pay machines to door actuators. In robots, however, QTouch chips could really shine. For example, Robosapien could have a couple of QT113 chips in its feet for sampling for the presence of water. Or, an actuator switch could be embedded in RS for receiving outside input without having to drill a hole in its plastic shell.

If you're looking for a painless way to experiment with the QT11x family of touch sensors, then look no further than Quantum's E11x QTouch Evaluation Board. Like so many other manufacturer evaluation boards, the E11x is loaded with lots of powerful features for allowing you to fully experiment with touch sensitivity sensors (Fig. 9-5). Both visual and audio touch indicators are included with the E11x and you can easily plug your own external electrode into the board. Oh, and an electrode is nothing more than a piece of metal foil or a loop of wire. All of this and more for less than $20 from Digi-Key.

FIGURE 9-5 **Unlike some evaluation boards, the Quantum E11x Evaluation Board can be dropped straight into Robosapien for adding some extra touch sensing capabilities for under $20. (Photograph courtesy of Quantum Research Group)**

CHAPTER 10

Message in a Bot

Are you a "remember when-er"? You know the type of person who will typically begin some boring account of a previously lost opportunity with the phrase, "I remember when...." While you'd love to reply, "Yeah, and monkeys can fly, too," instead you patiently sit there with your eyes glassed over, listening to every excruciating detail about this past glory. Unfortunately, these "remember when-ers" fail to realize that tomorrow is the date that they *should* remember, not yesterday. Why should they remember tomorrow? That's when the next great thing is going to happen. And you are (t)here.

This scenario always comes to my mind when I hear engineers wax lovingly about amplifier tubes, sliver halide film, magnetic recording tape, and Heath HERO robots. Get a life. In case you didn't read the office memo, transistors, digital cameras, ChipCorder®, and Robosapien are today's technology. OK, you've got a pretty good handle on most of this stuff, right? No? You're now asking, "What's with that ChipCorder thing?"

Think of ChipCorder as a digital, single-chip tape recorder that can provide high-quality playback without the need for battery backup circuits. Ahh, now I've got your attention, don't I? This incredibly simple chip requires only a handful of support components to record messages up to 40 seconds (e.g., for the I1600 Series) in duration and then play the message back on a standard 8-ohm speaker (Fig. 10-1). Additionally, the message has a claimed retention of 100 years, can be overwritten 10,000 times, and can cost less than $20 (depending on the exact ChipCorder series device). In fact, chances are that you might already have a ChipCorder in your house, right now. Many promotional key chains, picture frames, greeting cards, and stuffed toys that speak high-quality human phrases use ChipCorder devices for playback. Even some warning alarms in automobiles and industrial control centers rely on ChipCorder for belting out a clear and understandable signal. "Hey, stupid, STOP!" gets a person's attention far better than "Beep, beep."

FIGURE 10-1
A small sampling of the different types of ChipCorder ICs that are currently available. The newest ChipCorder, the I16xx series, is represented by the two small 16-pin ICs at the top of the antistatic foam pad.

The maker of ChipCorder, Winbond Electronics Corporation America, a wholly owned subsidiary of Winbond Electronics Corporation of Hsinchu, Taiwan, is based in San Jose, California. Started in 1990, the American branch began distributing signal-conditioning devices for consumer and industrial markets. With the 1998 acquisition of Storage Devices, Winbond America quickly became a market leader in silicon voice recording and playback integrated chip (IC) solutions—namely, ChipCorder. Additional acquisitions of Bright Micro Electronics and the thin film transistor (TFT) liquid crystal display (LCD) division of Cirrus Logic®, based in Austin, Texas, further entrenched Winbond America's contribution to award-winning voice and speech chip solutions and state-of-the-art TFT LCD-driver ICs. Four new product lines are aimed at bolstering Winbond America's speech products: the WTS70X series, the industry's first single chip IC solution that converts text to speech (TTS); the W68xx series, a family of voice CODEC chips aimed at telephony, communications, and consumer applications; the WMS72xx series, a family of 256-tap, nonvolatile, digitally programmable potentiometer ICs aimed at communications and industrial and consumer applications; and the ISD1600 ChipCorder series of single message record/playback ICs.

The ISD1600 is the first series of ChipCorder devices designed to operate from 2.4 to 5.5 volts (V). Furthermore, the ISD1600 series features 6.6 to 40 seconds in record/playback duration, push-button operation, LED indicators, nonvolatile message storage, and an integrated speaker driver, which provides both PWM and current-mode speaker outputs. By varying a user-determined external oscillator resistor, the ISD1600 series (1610, 1612, 1616, and 1620) can be programmed for a 4- to 12-kilohertz (kHz) sampling frequency. Likewise, this variable sampling frequency also determines the length of recording time. Finally, this is a fully integrated system-on-a-chip with support functions that include AGC, microphone preamplifier, speaker drivers, oscillator and memory. All of this in a neat 16-pin DIP. Perfect for hacking into a Robosapien.

In this hack, we are going to turn RS into a message center with an attitude. By inserting a ChipCorder ISD1610 device (with support components) inside Robosapien, family and friends can record short messages that can be played back on command. Now for playback you can either opt for the low-tech push-button method or go for broke and attach the ChipCorder playback function to one of the RS activity lines. In other words, you record your message on ChipCorder by pressing a button, and then you can hear your recording whenever you activate one of the Robosapien IR controls. For example, when a light beam is turned on in the left hand, your playback would begin. On the other hand (that's your hand, not Robosapien's), if you elect to switch to a different ChipCorder series, you can add several devices in series and make a variety of messages that can be individually played back. Just think, a different message for several different body movements, sensor triggers, or IR signals—and all of them are recorded in your voice. Now this hack is speaking my language. How about yours?

The Hack

STEP 1: A LITTLE OF THE NEW, A LITTLE OF THE OLD. Both the Robosapien speaker and the microphone will be used with this hack.

STEP 2: EXPRESS OR À LA CARTE. If you want to save yourself some time, you can purchase a ready-made ChipCorder recording board (Fig. 10-2). Just install this board, and RS will be ready to receive your commands. Otherwise, you can easily and inexpensively wire your own board (Fig. 10-3). Whichever circuit you choose, beware of electric motor noise. Shield the shoulder motors for reducing interference.

FIGURE 10-2
This ChipCorder experimenter's board can be dropped directly into Robosapien for the most rapid hacking turnaround time.

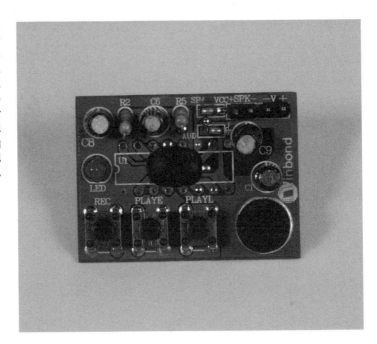

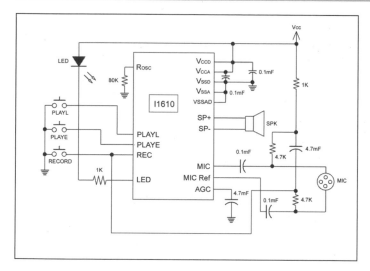

FIGURE 10-3
Schematic diagram of a simple ChipCorder digital recording and playback circuit.

STEP 3: ALL BUTTONED UP. At the least, you can build a ChipCorder circuit with only two push buttons—one playback (e.g., either PLAYL; level activated or PLAYE; edge activated) push button and one record push button. Or, you can add three push buttons. In this hack, I used a single edge-activated playback push button along with one record push button (Fig. 10-4).

FIGURE 10-4
Although you can barely see them, there are two micro push buttons mounted on the back neck of Robosapien for recording and playing digital recordings.

MSA Sound Allophone Component List

CODE	PHONEME	SAMPLE WORDS
128	IY	See, Even, Feed
129	IH	Sit, Fix, Pin
130	EY	Hair, Gate, Beige
131	EH	Met, Check, Red
132	AY	Hat, Fast, Fan
133	AX	Cotton
134	UX	Luck, Up, Uncle
135	OH	Hot, Clock, Fox
136	AW	Father, Fall
137	OW	Comb, Over, Hold
138	UH	Book, Could, Should
139	UW	Food, June
140	MM	Milk, Famous
141	NE	Nip, Danger, Thin
142	NO	No, Snow, On
143	NGE	Think, Ping
144	NGO	Hung, Song
145	LE	Lake, Alarm, Lapel
146	LO	Clock, Plus, Hello
147	WW	Wool, Sweat
148	RR	Ray, Brain, Over
149	IYRR	Clear, Hear, Year
150	EYRR	Hair, Stair, Repair
151	AXRR	Fir, Bird, Burn
152	AWRR	Part, Farm, Yarn
153	OWRR	Corn, Four, Your
154	EYIY	Gate, Ate, Ray
155	OHIY	Mice, Fight, White
156	OWIY	Boy, Toy, Voice
157	OHIH	Sky, Five, I
158	IYEH	Yes, Yarn, Million
159	EHLL	Saddle, Angle, Spell
160	IYUW	Cute, Few
161	AXUW	Brown, Clown, Thousand

162	IHWW	Two, New, Zoo
163	AYWW	Our, Ouch, Owl
164	OWWW	Go, Hello, Snow
165	JH	Dodge, Jet, Savage
166	VV	Vest, Even
167	ZZ	Zoo, Zap
168	ZH	Azure, Treasure
169	DH	There, That, This
170	BE	Bear, Bird, Bead
171	BO	Bone, Book, Brown
172	EB	Cab, Crib, Web
173	OB	Bob, Sub, Tub
174	DE	Deep, Date, Divide
175	DO	Do, Dust, Dog
176	ED	Could, Bird
177	OD	Bud, Food
178	GE	Get, Gate, Guest
179	GO	Got, Glue, Goo
180	EG	Peg, Wig
181	OG	Dog, Peg
182	CH	Church, Feature, March
183	HE	Help, Hand, Hair
184	HO	Hoe, Hot, Hug
185	WH	Who, Whale, White
186	FF	Food, Effort, Off
187	SE	See, Vest, Plus
188	SO	So, Sweat
189	SH	Ship, Fiction, Leash
190	TH	Thin, Month
191	TT	Part, Little, Sit
192	TU	To, Talk, Ten
193	TS	Parts, Costs, Robots
194	KE	Can't, Clown, Key
195	KO	Comb, Quick, Fox
196	EK	Speak, Task
197	OK	Book, Took, October
198	PE	People, Computer
199	PO	Paw, Copy

STEP 4: I CAN'T HEAR YOU. Wire the ChipCorder circuit into the Robosapien speaker.

STEP 5: ROBOSAPIEN, CAN YOU HEAR ME? The Robosapien will be used for the MIC input on the ChipCorder. Two discrete connections must be made to this input. Due to excessive glue on the Robosapien main circuit board, I found that it is best to gain access to this microphone from the backside of this board.

STEP 6: 1-2-3, CHEQUE. Attach a battery pack to the ChipCorder circuit. Optionally, if you are able to locate the new I1610 IC, you can drive the circuit straight off of the Robosapien main circuit board. Install batteries in Robosapien and make a sample recording. Now playback your recording. Robosapien never sounded so good, right?

Ouch, Something's Wrong

Although there aren't very many components in a ChipCorder circuit, the wiring can be somewhat daunting. Double and triple check your wiring. If you think that the Robosapien speaker and/or microphone connections are at fault, attach a spare 8-ohm speaker to the ChipCorder circuit's output for checking the speaker connection and a spare electret microphone to the MIC input for verifying the microphone connection. Finally, use your multimeter to make sure that the ChipCorder IC is receiving ample power.

Rosebud..., One More Thing

If you'd rather opt for a voice other than your own, Magnevation LLC of Capshaw, Alabama has just the ticket for you to ride. SpeakJet™ is a single-chip speech synthesizer and complex sound generator using Mathematical Sound Architecture™ (MSA) that can be tied to an OOPic or BASIC Stamp I/O line (Fig. 10-5). Or, you can directly address the SpeakJet via a push-button interface. Prerecorded with 72 allophones, 45 sound effects, and 12 DTMF tones (e.g., the tones emitted by a push-button telephone), SpeakJet is the ideal solution for implanting a unique set of vocal cords in Robosapien.

In a remarkable nod to robot builders, Magnevation has included two RC input control pins on SpeakJet (Fig. 10-6). These inputs can be triggered via a variable pulse width signal from a conventional RC model receiver for playing up to four precanned phrases, sound effects, or control functions.

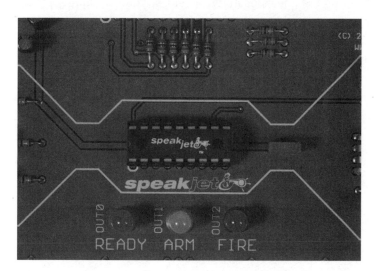

FIGURE 10-5
SpeakJet from Magnevation is a complete speech synthesizer package that can be easily integrated into Robosapien.

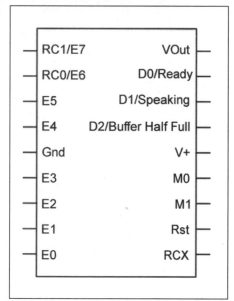

FIGURE 10-6
SpeakJet can be controlled by many of today's commonly used microcontrollers.

RC1/E7	VOut
RC0/E6	D0/Ready
E5	D1/Speaking
E4	D2/Buffer Half Full
Gnd	V+
E3	M0
E2	M1
E1	Rst
E0	RCX

For such a capable speech synthesizer, the power requirements for Speak-Jet are certainly modest. A supply voltage between 2.0 to 5.5 volts direct current (VDC) is needed for powering the SpeakJet. In fact, this power could be pulled directly from the Robosapien motherboard.

Even better, by using the SpeakJet Activity Center, you can rapidly prototype an entire speech synthesis application with programming control supplied by an OOPic or BASIC Stamp (Fig. 10-7). The subsequent playback of your application inside Robosapien can then be reduced to the microcontroller of your choice, a power supply, and speaker. That's it.

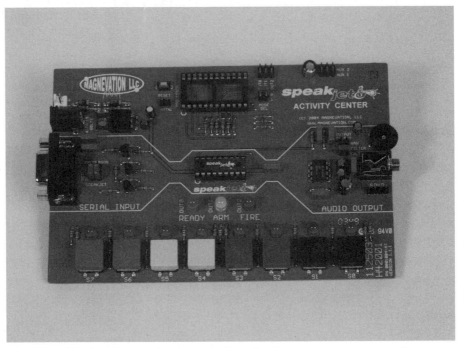

FIGURE 10-7 The SpeakJet Activity Center is a simple way to create a speech synthesis application for usage in Robosapien.

CHAPTER 11

Twin Robot Brothers of a Human Father

Did you ever see the 1972 MGM movie *The Thing with Two Heads*? This movie was about a bigot doctor played by Ray Milland, who has his head grafted onto the body of a death row inmate played by Rosie Grier. Not to be confused with the 1971 MGM flick *The Incredible Two-Headed Transplant*, in which the head of a killer played by Albert Cole is grafted onto the body of a caretaker's mentally handicapped son, played by John Bloom.

The trailer copy for *The Thing with Two Heads* read something like this: The most fantastic medical experiment ever dared! They transplanted a white bigot's head onto a soul brother's body! And now, with the fights and the fuzz, the choppers and the chicks, they're in deeeeep trouble!

Confused? While Hollywood must've thought that two heads were better than none in these twin low-budget, campy sci-fi flicks, there is an interesting segue into a powerful Robosapien hack here. Please bear with me.

What if you were to graft two Robosapiens together? Not in some cheesy, amateurish Hollywood way, but in a real attempt at improving on the original solo RS design. You know the drill: Make it faster, stronger, better—cue *The Six Million Dollar Man* theme song. Since Lee Majors isn't at my disposal, let's double our pleasure with a siamese Robosapien, instead. Unlike these Hollywood thrillers, my "thing with two heads" will be an upper-chest front to upper-back joining.

Technically speaking, this tandem arrangement must be in complete synchronization with each movement—you wouldn't want a left foot of one Robosapien tripping over the right foot of the other one. Furthermore, you don't really need two heads, do you? Think of all of the conflicting infrared (IR) signals that would have to be sorted out, interpreted, and then sent to the correct robot. Ouch. Finally, the connector between these siamese-twin Robosapiens must be long enough and flexible enough to permit full travel of both the arms and the legs. It wouldn't hurt either, to design a connector with adequate vertical flexibility to allow one RS to be (temporarily) at a slightly different elevation than the other Robosapien. Think of this connector as a moveable boot that can join the two robots together, encase all of the needed hardware for this hack, and enable the Robosapiens to function as a unit rather than two tethered individuals.

In fact, the trailer copy of this hack reads something like this: The most fantastic Robosapien hack ever dared! They transplanted the rear end of Robosapien body onto the front end of a Robosapien's body! And now, with the fights and the farts, the choppin' and the hoofin', they're in deeeeep trouble!

The Hack

STEP 1: YOU LEAD, I'LL FOLLOW. Decide which RS will lead and which one will follow, then remove the chest back plate from the leader and the chest front plate from the follower.

STEP 2: MAKE ROOM FOR DADDY. Inside the chest back plate, remove the speaker cover and pry the speaker out of its socket (Fig. 11-1). In every Robosapien that I've hacked, this speaker is glued in place very tightly. A flathead screwdriver can be used for pry the speaker up and out, but don't mar the plastic ring by leveraging the screwdriver against it.

STEP 3: DRILL TO KILL. A conventional woodworking hole saw is ideal for drilling out the plastic from both chest plates (Fig. 11-2). Remember to use *very* slow speeds, or the plastic will melt and could break. My flexible conduit

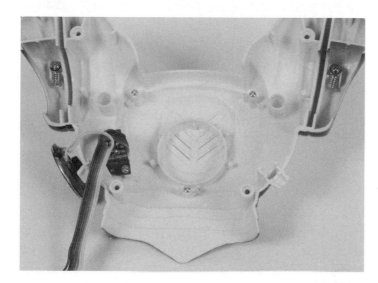

FIGURE 11-1
Remove the speaker from the master Robosapien's back plate.

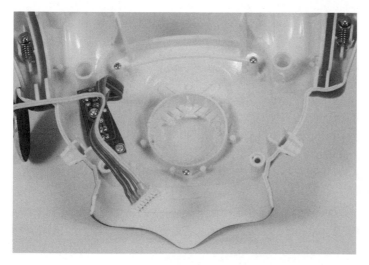

FIGURE 11-2
Carefully drill a hole in the back plate. Go real slow with your power hole drill, or you might ruin the plastic.

was 1 3/16 inches in diameter. Therefore, I used a hole saw that was 1 1/8 inches in diameter. This slightly smaller diameter allowed me to create a snug fit and hide my drill marks. The result was a clean and professional-looking installation (Fig. 11-3).

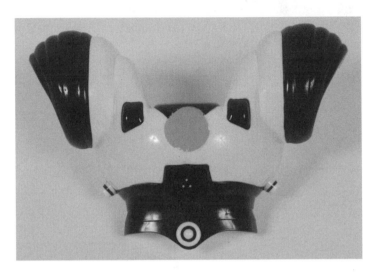

FIGURE 11-3
A second hole is drilled in the Robosapien slave's front plate.

STEP 4: BE GENTLE, DOC. Insert the conduit into both the front and back plates (Figs. 11-4 and 11-5). If your fit isn't snug enough, you can glue the conduit in place with plastic cement (Fig. 11-6). Ensure that both the front plate, the conduit, and the back plate are properly aligned before you apply any glue.

STEP 5: ONE HEAD IS BETTER THAN TWO. The greatest mistake you can make is attempting to drive your new Robopede with two brains and two IR receivers. Although they may synchronize for a minute or so, they will eventually become unsynchronized, and the entire hack could get damaged. A better solution is to use the front Robosapien as the "master" and the following Robosapien as the "slave." In this context, the slave's IR sensor is removed, all motor connections are piggybacked onto the master's main circuit board, and the slave's speaker is connected to the master's sound output. Basically, you can remove the slave's main circuit board and let the master run both robots.

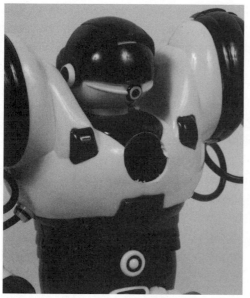

FIGURE 11-4 I am your slave. The slave Robosapien is ready to receive the plastic conduit.

FIGURE 11-5 And I am your master. The master Robosapien will now drive the speaker inside the slave Robosapien.

FIGURE 11-6 The plastic conduit is test fitted in the chest of the slave Robosapien. Oh, and I still giggle every time I see this photograph, too.

I did encounter some problems with this simplified approach, however. The power output from the master RS is significantly drained by this tandem team. Therefore, you should consider limiting the slave's motor control to just the hip servos. Alternatively, you could retain the slave's IR receiver and main circuit board while routing the hip servo motor control to the master's main circuit board. This variation would enable you to use one IR remote control for sending signals to both the master and the slave RS, while the master Robosapien would be solely responsible for controlling all walking movements. Either way, use the conduit for holding all wiring between the two Robosapiens. Now seal 'em up, power 'em up, and walk 'em around a bit (Fig. 11-7).

FIGURE 11-7
A Robopede. Make sure that you leave enough gap between the two Robosapiens so they don't accidentally hit each other while moving.

Ouch, Something's Wrong

Properly implementing the master/slave control system is vital for success. The easiest method for ensuring that the front Robosapien controls the rear Robosapien is to remove the rear Robosapien's main circuit board. Then all motor and sensor control *must* be routed to the master RS. Once you've gained some experience from this single-brained approach, you can enhance your hack by adding the second main circuit board back into the slave RS.

Rosebud..., One More Thing

Hey, you know that bionic six-million-dollar Robosapien idea isn't a bad follow-up to our homage to Ray Milland and Rosie Grier. Now that you've got a siamese Robosapien, let's soup it up a bit. Of course, you can apply these advanced hacks to a singular Robosapien, as well.

It's possible to triple the walking speed of an RS by using rechargeable batteries, some FET H-bridge post buffers, and two radio-control-grade high-torque, high-RPM motors in the hip gearboxes (Fig. 11-8). The second "running" forward walk mode will now give the robot more than enough speed for robot soccer applications and kicking the tar out of your next door neighbor.

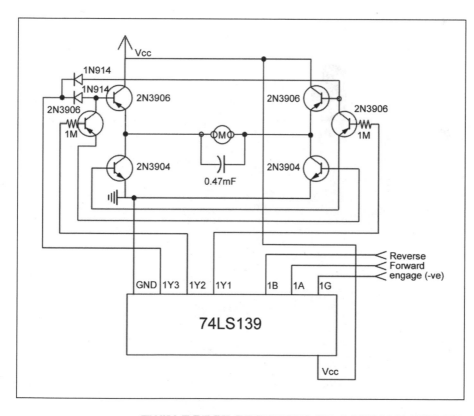

FIGURE 11-8 Schematic diagram of an H-Bridge circuit based on a 74LS139. (Schematic redrawn from original design by Mark W. Tilden; © 1997)

Another notion is reusing the controller from one of the twin Robosapiens in another robot. Many of the WowWee prototypes were originally controlled by one or two RS brains and brawn. As a universal controller, it's small, cheap, robust, and modularly programmable, making it easy to pop into different bodies and shells.

A little known trick for souping up RS is that the processor crystal (Y1) can be swapped out with a ceramic resonator capacitor, for example, and that the frequency of the robot's operation can be slowed down or sped up by almost 50% just by using different sized caps (Fig. 11-9). Plus the remote will still work. This allows you to vary the speed of motor actions considerably for fast/lightweight or slow/powerful designs. Additionally, electrically shorting this crystal also can freeze the bot for long periods without suffering time out.

Finally, remember different bodies, same ol' brain. Don't forget the brain is the stumbling block in Hollywood, only. Don't sweat it, just hack it.

FIGURE 11-9 Replacing the capacitor crystal (Y1) on the main circuit board can dramatically affect the speed of the Robosapien motor actions. In this case, a ceramic resonator capacitor is going to be soldered onto the Robosapien main circuit board.

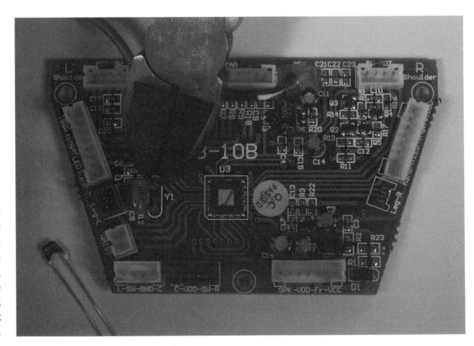

One Is Better Than Two

History is littered with examples of two (or more) products being combined into one big product. This zeal for gluing stuff together reached its zenith during World War II.

Sure, the German Luftwaffe was able to crank out weird doppelgängers on a regular basis—when you're losing, you get desperate. But who would have thought that the U.S. Army Air Corps (later the U.S. Army Air Force, or USAAF) would entertain such a notion? Even odder, the focus of this USAAF design grafting was the sensational North American P-51 Mustang.

Originally dubbed the P-82 Twin Mustang, this aircraft was supposed to be the Air Force's solution for long-range escort fighters flying long 8-hour missions over the Pacific Ocean. Having two pilots inside the same aircraft for sharing navigational duties seemed like a good idea. So, North American (the aircraft's manufacturer; which later became Rockwell, which later became Boeing) settled on gluing two P-51 Mustangs together along the wing(s) and the horizontal stabilizer(s) with the "pilot" being seated in the left fuselage and the "copilot" being housed in the right fuselage (Fig. 11-10).

Construction was started in 1944, with the first flight on June 16, 1945. Rats! The war was over, so the ordered construction of 500 F-82B (the "pursuit" P was dropped in favor of the "fighter" F designator) was reduced to an order of 20 and, then later, increased to over 200 production models.

Radical improvements in airborne radar systems, however, changed both the shape and the designation of the Twin Mustang.

FIGURE 11-10 Before Robopede, there was the North American F-82B Twin Mustang.

Now called the F-82E day escort fighter and the F-82F and F-82G night fighters, these twins each featured a central radar pod that was mounted under the middle wing section. Furthermore, the "right-hand" copilot was now tasked as the radar operator.

Believe it or not, another war was started, and the Twin Mustang was thrown into service. In the Korean War, three significant events swirled around the F-82. First, the USAAF was renamed the U.S. Air Force (USAF). Second, the F-82 became the first USAF fighter aircraft to be deployed to the Korean War. And, finally, an F-82G piloted by Lt. William Hudson scored the first U.S. aerial kill of the war.

At the end of the war, 14 F-82F and G were assigned cold weather fighter interception duty in Alaska. This reassignment mandated the new designation of F-82H for these Twin Mustangs. Today you can find examples of Twin Mustangs at the National Museum of the USAF in Dayton, Ohio.

TWIN MUSTANG DATA
(All data from USAF Museum Archives)

SPECIFICATIONS

SPAN: 51 feet, 3 inches

LENGTH: 38 feet, 1 inch

HEIGHT: 13 feet, 8 inches

WEIGHT: 24,800 pounds maximum

ARMAMENT: Six .50-caliber machine guns

ENGINES: Two Packard V-1650s of 1,380 horsepower each

COST: $228,000

SERIAL NUMBER: 44-65168, on display in Museum

PERFORMANCE

MAXIMUM SPEED: 482 miles per hour (mph)

CRUISING SPEED: 280 mph

RANGE: 2,200 miles

SERVICE CEILING: 39,000 feet

CHAPTER 12

If I Only Had a Brain

While you might think that Ray Bolger's Scarecrow quest for intellect from the 1939 movie *The Wizard of Oz* is a suitable goal for your final RS hack, you probably haven't even begun to explore the true potential that is housed within the onboard Robosapien brain. You see, Robosapien is both brawn and *brains*. In fact, the controller card can be easily removed from Robosapien and transplanted into a wide variety of other robots and devices. Just look at Chapter 8 for some great inspiration for doing just this type of hack.

If you insist on giving Robosapien a lobotomy, there are several great candidates for inserting into the RS noggin for supplying the needed controlling processes. Presented in alphabetical order, is my heavily edited short list of the best in robot brains.

BASIC Stamp 2

In 1992, Parallax (www.parallax.com) kick-started the fledging DIY robot movement with the release of the BASIC Stamp Rev. D. By the end of 1998, Parallax had sold over 125,000 of these BASIC Stamp modules. Today the BASIC Stamp microprocessor line includes eight different types of BASIC Stamp modules with varying capabilities. For transplantation into Robosapien, the best Parallax candidate is the ubiquitous BASIC Stamp 2 module.

This 24-pin DIP module includes an onboard processor, memory, clock, and interface (via 16 I/O pins). Controlled by a derivative noncompiled BASIC programming language known as Parallax BASIC (PBASIC), which contains 42 commands, the BASIC Stamp 2 is able to monitor and control motors, timers, switches, sensors, relays, and valves via simple access to its I/O lines (Fig. 12-1).

FIGURE 12-1
The venerable BASIC Stamp 2.

Programming the BASIC Stamp 2 module is best achieved via the Parallax Board of Education® (BOE). In either a serial or a USB interface configuration, the Parallax BOE supplies a rudimentary platform for gaining access to every aspect of the BASIC Stamp 2 module. (*Note:* The BOE is just about too big to fit inside the Robosapien chest cavity. If you're game, you can trim the corners and carefully wedge it into place.) The programming of the BASIC Stamp 2 module on a BOE can be done either on a Mac (with the neat program MacBS2 by the inventive Murat N. Konar; www.muratnkonar.com/otherstuff/macbs2/index.shtml) or a PC.

BASICX-24™

Another all-in-one BASIC programming module is from NetMedia (www.basicx.com). Packaged in a similar 24-pin DIP, the BASICX-24™ houses 21 user-programmable I/O lines, onboard status light-emitting diodes (LEDs), 32-kilobyte (kb) EEPROM, 400 bytes of RAM, and a serial/parallel programming interface (Fig. 12-2).

FIGURE 12-2
This BASICX-24 is actually the brain that is shipped with Botster.

One strong plus for the BASICX-24 module is the presence of a robust BASIC programming language implementation. The BASICX language features 80+ commands, including seven valuable string functions should you wish to add a liquid crystal display (LCD) screen to your Robosapien. This depth in programming capability enables the development of powerful applications for enabling RS to perform data logging, access ultrasonic range finder measurements via a Polaroid® sonic sensor, and control more powerful RC servos. Also, its performance specs make the BASICX-24 a screamer—able to execute 65,000 BASIC instructions per second.

Not being the "big" guy on the block does have its cost, however. To program the BASIC X-24 module, you will need the BASICX-24 Development Kit (Fig. 12-3). Shipped as a 3¼-inch × 3½-inch board that has ample room for adding the included BASICX-24 module and an optional 2 × 16 serial LCD module, the board also supplies four user-programmable micro push buttons, and this board is tough to squeeze into RS. The real knock against the BASICX-24 module, however, is that programming in BASICX is only supported on PC platforms.

FIGURE 12-3
The BASICX-24 Development Kit.

BrainStem® GP 1.0

For about 10 years there was only one name in home-brew robotics: Acroname (www.acroname.com). Since 1994, this small Boulder, Colorado, company has worked very hard at supplying quality, hard-to-find parts at affordable prices to robot builders, hobbyists, and experimenters. Their most popular microcontroller is the epitome of this working class ethic.

The BrainStem General Purpose module (GP 1.0) supports five 10-bit A/D inputs, five flexible digital outputs (i.e., for a combined 10 I/O lines), a Sharp GP2D02 driver port, an IIC bus, and four high-resolution servo outputs (Fig. 12-4). Measuring a scant 2 1/2 inches square, the BrainStem GP 1.0 module is a good candidate for fitting comfortably inside Robosapien.

FIGURE 12-4
Acroname's BrainStem GP 1.0.

A module can operate in three different modes, simultaneously, and stacked with other BrainStem modules and/or other manufacturer's devices on its 1-Mbit IIC bus. These three operation modes include use as a slave device, execution of up to four concurrent tiny embedded application (TEA) programs (although BrainStem can store up to 11 1-kb TEA programs), and reflex actions.

When using a BrainStem command set, modules can be operated in a slave mode. These are simple direct control operations that are typically sent via the Acroname Console application. Other avenues for achieving a similar slave mode can be controlled through configuration files, serial IIC bus devices, TEA programs, and reflex actions.

The TEA language is a subset of the C programming language and is compiled, loaded, and run via the Acroname Console application. Console includes a little embedded application fragment (LEAF) compiler that is used for manipulating BrainStem reflex operations.

There aren't a lot of frills with TEA programs. Programs are precompiled, however. So you can employ conditional compilation, add macros, or include other files. In normal operation, TEA programs are very small, under 1 kb in size, and have no memory allocation, structures, or objects. All variables are stack based; the stack can be very small on some environments so minimal recursion is possible.

In the last operational mode, reflex actions, you can set up automatic responses based on input variables. For example, stopping a robot's motors when a touch sensor is depressed. Reflex actions manage the mundane control (e.g., turning motors on/off), monitoring (e.g., watch battery voltage levels), and reaction functions (e.g., closed-loop controls) within the BrainStem module.

Finally, Acroname has continued its tradition of making its products accessible to the widest possible audience of robot builders. All of its software and applications are built for Linux, Mac (including 8.x, 9.x, and OS X), Palm, and Windows operating systems. This is a refreshing distribution plan when you realize that many other manufacturers blindly suffer from PC envy.

Handy Cricket

Is there any robot builder who doesn't know about the Handy Board? The Handy Board is a 68HC11-based microcontroller board designed by the Massachusetts Institute of Technology (MIT). In a remarkable gesture to the robotics community, MIT has licensed the Handy Board design as freeware for educational, research, and industrial use. While you can readily download design information about building your own Handy Board on the Web (www.handyboard.com), there are a couple of manufacturers who sell completed Handy Boards. Rising above all of this clutter is Gleason Research (www.gleasonresearch.com). Since 1995, Gleason Research has been selling the MIT Handy Board to robot builders worldwide. Now, a smaller freeware design is available, and it is perfect for installation inside Robosapien.

The Handy Cricket Version 1.1 is a low-cost module based on the Microchip PIC® microprocessor featuring a built-in Logo interpreter (Fig. 12-5).

FIGURE 12-5
Handy Cricket is handy indeed—for robot hacking.

Equipped with two motor ports, two sensor ports, two bus ports, 4 kb of static memory, and a piezo speaker, the Handy Cricket connects with the host PC via a serial port IR Interface Cricket (Fig. 12-6). This infrared (IR) interface is just like the IR Tower that most of you are familiar with from the LEGO RIS 2.0. One of the exciting attributes of the Handy Cricket, however, is the action that contributed to its name.

A unique IR transmitter/receiver circuit built into the Handy Cricket enables communication between two or more Handy Crickets. Like the chirping of a cricket, the Handy Cricket is able to "chirp" IR signals at a 50K data

FIGURE 12-6
An IR communication link is used for connecting the Handy Cricket to a computer.

rate between each other. Imagine this—you could have various Robosapien Handy Cricket brain transplants "talking" between each other. Oh, and just like a biological cricket, the Handy Cricket is a tiny sucker. The overall dimensions are just a bit under 2½ inches per side (Fig. 12-7).

Sure, all of this hardware stuff is exciting, but the part of the Handy Cricket that leaves me salivating is the fantastic implementation of the Logo programming language, called Cricket Logo, that is built into the Handy Cricket.

You can have your BASIC modules and your C modules, if you want, but if you really want to program a robot, there is hardly a better "tongue" to use than Logo. Derived from a turtle-based language used in the late 1980s for introducing school-age children to programming, Logo became hot, then died quickly from BASIC BS.

The Handy Cricket is programmed in a language called Cricket Logo, which is a simplified version of the powerful yet easy-to-learn Logo language. Unlike most programming languages, Cricket Logo (as well as its ancestral Logo) is

FIGURE 12-7
**A variety
of optional
interfaces can
be plugged
into the Handy
Cricket.**

short, simple, sweet, and primitive. I'll grant you, Logo is a little archaic, but the payback can be terrific for robot builders. For example, you want to control two motors, then you might need to learn 12 simple commands, or six elementary commands for interacting with two sensors. Likewise, control functions are only a couple of lines worth of code, and commonsense commands like BRAKE and SETPOWER don't require a lot of human smarts to figure out.

Inside Cricket Logo, you get a 16-bit number system, global variables, timing functions, and data recording/playback primitives. If you really want to crack the code for the Handy Cricket, however, the SEND and IR primitives for two-way IR communication are worth your every penny. Like any responsible robot manufacturer, however, the best part about the Handy Cricket is that Gleason Research has made Cricket Logo available in both Mac and Windows operating systems. Hey, including a simple little robot car plan complete with an eight-line Cricket Logo program on the last page of the user's manual, doesn't hurt, either.

TiniPod™

OK, if you want smaller, then the TiniPod from New Micros of Dallas, Texas, can be inserted into Robosapien with room to spare (Fig. 12-8). But this type of manufacturing achievement is nothing new to New Micros. Although founded more than 20 years ago, in 1987, New Micros set the trend for today's robot all-in-one microprocessor modules with a stand-alone, single-chip computer with a built-in programming language. The release of the slightly more than 1-inch-square TiniPod marks a pinnacle in robot controller development.

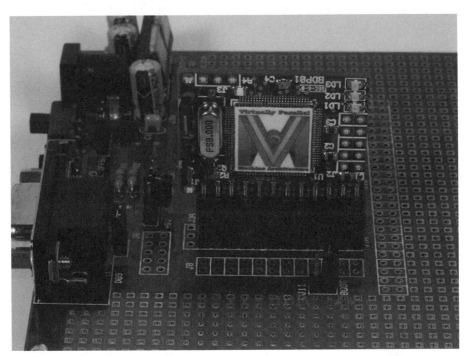

FIGURE 12-8
TiniPod from New Micros.

Like its contemporaries, TiniPod sports 16 digital I/O lines that can drive servos and timers; a DSP56F803 MPU, 16-bit processor; a 32-kb × 16-bit Program Flash EEPROM; and 2-kb × 16-bit RAM for data. Unlike other controller modules, TiniPod includes a high-level language known a IsoMax™.

Even though IsoMax has an English-like command set, understanding and fully exploiting its intricacies can be difficult. So be forewarned. IsoMax relies on state-machine programming techniques for real time tasks. This state machine is known as a virtual machine. Therefore, programming in IsoMax becomes a description of a virtual machine that gathers environmental information and then (re)acts according to predefined actions and enters into a new state. You can also liken these states to threads. These states, threads, or virtual machines can be strung together inside TiniPod's memory.

In a nod to multitasking, IsoMax is able to run these multiple virtual machines simultaneously. This forms an interactive environment where threads can run in the background, independent of each other. New Micros indicates that this performance is executed in a parallel fashion, called a Virtually Parallel Machine Architecture (VPMA).

Because IsoMax is a case-sensitive language, and unlike many other robot controller module high-level languages, an example is important here. For example, consider this fake code snippet:

```
STATE-MACHINE xxx

        ON-MACHINE xxx

                APPEND-STATE yyy

                APPEND-STATE zzz

IN-STATE yyy

        CONDITION test @ LOW-LIMIT @ i

        CAUSES action

        THEN-STATE zzz

        TO-HAPPEN

IN-STATE zzz

        CONDITION test @ LOW-LIMIT @ i

        CAUSES action

        THEN-STATE zzz

        TO-HAPPEN
```

Ow, that's tough. This really bad example gives you just a rudimentary idea about programming with IsoMax. IsoMax on TiniPod is available for Windows. Alternatively, programming environments in Small C, Assembler, and, my personal favorite, FORTH can be optionally purchased and licensed.

No matter which high-level controller module you select for transplantation into Robosapien, just make sure that you do your hack with some style. One amateur hacker posted his abomination mutation of a Robosapien with an Atmel microcontroller stuck in the poor robot's back. You can do better, and if you follow my lead, I'm sure that you will do better.

The Hack

There are no steps in this type of hack—either it works or it doesn't. Begin by removing the Robosapien main circuit board (Fig. 12-9). Take your chosen "brain implant" and hook it up to as many of the Robosapien motors and sen-

**FIGURE 12-9
Ready for receiving a new brain implant.**

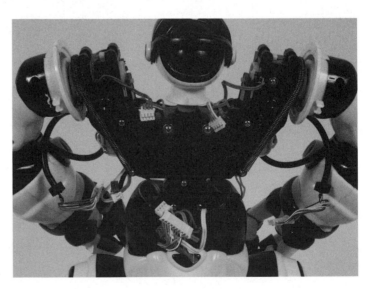

sors as possible (Fig. 12-10). Make sure that you observe the power requirements of both your brain implant *and* Robosapien. Unless you have a really unique brain implant, be prepared to snip apart all of the Robosapien motor and sensor plugs for connection to your microcontroller's appropriate ports.

FIGURE 12-10
The Handy Cricket is being test fitted in the Robosapien brain pan.

Ouch, Something's Wrong

Two of the most common faults with a brain implant are a power voltage incompatibility and a wiring error. In the case of the power problem, make sure that your selected microcontroller does not require *more* than 3.5 volts (V). If it does, add additional batteries for powering the brain implant. With regard to a wiring snafu, remember this: The more onboard functions that you can implement with your microcontroller, the better. For example, the Handy Cricket has an attached power supply, onboard speaker, *and* its own IR communication port. With this kind of rich feature set, you can't go wrong. Though I did have a lot of difficulty getting two of the touch sensors to work properly.

I've Fallen, But I *Can* Get Up

Are you amazed with Robosapien's ability to walk? I sure am; even more so, when I study the brilliant use of springs for conserving valuable motor energy, as well as helping to keep the bipedal lug balanced. If you had a budget larger than the sub-$100 cost for a Robosapien, you might think about installing an accelerometer for maintaining the balance of your Robosapien.

If this proposition sounds intriguing, then you should study the accelerometers offered by MEMSIC. These CMOS chips rely on a gaseous proof mass for measuring internal changes in heat transfer that is caused by acceleration. Unlike other accelerometers which use a solid proof mass, these MEMSIC chips are able to handle greater shocks and don't suffer from particle contamination. Huh? Are your eyes glassed over yet? Suffice it to say that a heat source is suspended inside the chip over an opening, which is surrounded by four thermopiles. These thermopiles output a voltage change as IC moves in the direction of a tilt. Better now?

Is this too exotic for your taste? Parallax doesn't think so; their Toddler robot uses a BASIC Stamp for reading and reacting to tilt via a MEMSIC accelerometer. If you'd like to play with a simple accelerometer from MEMSIC, then the $25 LED Tilt Motion Demonstration Board is for you (Fig. 12-11).

This demo board gives you a quick visual reference for tilt, shock, and acceleration. Just put this board in your palm and wiggle your hand. The sensor will read your hand's orientation and motion, and it will display the results on the board's LEDs. Now when you applaud at the theater, everyone will be able to see your appreciation, as well as hear it.

FIGURE 12-11 **One tilt in any direction will send this MEMSIC tilt demo board into a flurry of LED activity.**

Rosebud…, One More Thing

Yes, I used the Handy Cricket as my Robosapien implant—mainly because I'm a sucker for Logo. I would love to have used the TiniPod, but I just couldn't get a good solid handle on IsoMax, and I didn't have any budget for buying a FORTH license. Similarly, any of the BASIC modules or BrainStem are great candidates for adding some brain muscle to RS. In these cases, however, you will need to purchase a programming interface or development kit for loading your code into your chosen module.

If you're really adventuresome, and if you're a hacker, you might consider using one of the hundreds of microcontroller chips that are currently available. Atmel and Microchip are the big guns in this arena, but neither of these manufacturers wanted to assist with the preparation of this portion of the book!

Epilogue

Silly love songs aside, there really can be love at first sight. I should know, it happened to me on November 26, 2004.

You remember *that* day, don't you? That's was the day right after Thanksgiving Day. The much hyped "most busy shopping day of the impending holiday season." Having to kill some time while I waited for an upcoming family rendezvous from a postshopping spending spree, I wandered the aisles of my favorite haunt—the local toy store.

No matter where you live in the world, you can always find some great robot building and hacking inspiration in your local toy store. From trends to parts and from subjects to castoffs, the toy store can be a terrific source for all of your robotic research needs.

It was somewhere along the third aisle, the aisle that hawks remote-controlled (RC) cars and trucks, that I turned and met eye to LED (light-emitting diode) the perfect robot toy. Robosapien, from WowWee Ltd., screamed innovation and consumed me with inspiration for hundreds of robot projects that I had only previously dreamed about. Mobile wireless video cameras, radio-controlled alarm clocks, siamese robots, you name it, I was thinking about Robosapien as my testbed for each project.

Remarkably enough, I didn't have long to dream about potential robot hacking opportunities, startling quick hands were grabbing my newfound robot friend off of the shelf like hotcakes right under my nose. I kid you not, I literally held onto the last Robosapien as an irate parent attempted to pry it out of my tight little fingers. Mustering my best lunatic look I won parental rights and that parent settled on a vastly inferior RC Hummer. I walked out of the toy store victorious.

I had to play this smart, though. A toy for Dada; I could hear it now, "You've got to be kidding?" So with the Prochnow herd within sight, I made a beeline to the van and tucked Robosapien inside the back hatch. Later, on Christmas Day I was delighted to see that Santa Claus had left me a brand new Robosapien. And later that week, my first Robosapien hack was born.

Waddling into the living room, Robosapien transmitted a video signal of the kids playing with their new toys to me, at my computer, in the next room. Well, actually, that was my third hack. I had already installed some rechargeable batteries in Robosapien *and* programmed an infrared (IR) transmitter for signaling command sequences from one room into a different room. I was bitten by the Robosapien hacking bug.

Over the next several months, dozens of Robosapiens were opened, disassembled, modified, reassembled, and tested. These were my dream hacks. Hacks that I had fantasized about on that fateful day after Thanksgiving 2004. Some were successful, some were ridiculous failures, but, most important, Robosapien was a constant, a stable platform, that enabled each and every hack to be evaluated on equal footing (Fig. E-1).

Packaging all of my successful hacks into a tidy package, the entire manuscript was presented to McGraw-Hill, which, with the wisdom that only comes from a market-leading publisher, created the book that you are now holding in your hands. Along the line, however, a once-in-a-lifetime acquaintance was made. Through the deft guidance of WowWee and McGraw-Hill, I was able to make contact with the remarkable designer of Robosapien—Mark W. Tilden.

No matter what you've read about Mark Tilden, or heard about him, believe me when I tell you that he is one of the most generous and creative technologists that I have ever met. Oh, and I've met some of the big boys (and big girls, too) in today's world of high-tech gizmos, gadgets, and toys. He is, however, the real deal. Friendly, helpful, intellectual, and, lucky for me, sarcastic to a fault, Mark fired back a reply to my initial electronic query from halfway around the world.

Dazzling me with his comprehensive reply, it wasn't until I read his signature tag line that I knew that Robosapien was 100% Mark W. Tilden (i.e., you

know, "a fusion of technology and personality"). In that message-concluding line Mark had parenthetically inserted the statement "sent from my phone," meaning that, his entire verbose reply had been clanked out on his BlackBerry cell phone PDA (personal digital assistant). Good God, I fired back, I won't bother you again until you're able to get to a computer, so that you won't have to "fat finger" in your messages to me. Mark instantly shot back, I always fat finger everything, so send me your questions. I did and you have just bene-fited from that encounter.

Dave Prochnow
(not from my phone)

Sharply Dressed Sapien

Congratulations; here is your bonus 12th hack.

In a *big* departure from the hacks found earlier in this book, this final hack will test an entirely different set of your creative talents.

There's nothing worse than a naked Robosapien. Its cold, harsh, plastic body is just so—I don't know—robotlike. So let's sew RS a new stylish robe. You know, something that a robot can throw on right after a shower, a quick cover-up for down by the swimming pool, or a smoking jacket for leisurely sitting around the robot workshop waiting for the next sizzle of solder. OK, smart aleck, I'm trying to give you an option for hiding all of those "wrong" holes that you drilled in its chest for holding the wireless video camera lens from the hack in Chapter 4. OK?

The Hack

STEP 1: LAID OUT IN PLAID. Cut out the necessary pattern pieces, lay them out according to the instructions packaged with the pattern, and cut out each fabric piece (Figs. A-2 and A-3).

STEP 2: SEW WHAT? Sew the front to the back at the shoulders. Fold and press the collar. Stitch the facings along the back seam, and sew the sleeves in place. To accommodate the exterior wiring of Robosapien, leave the side seams open. You should finish these seams, however, to prevent fraying of the fabric.

STEP 3: A HEM. Sew a hem along the bottom of the robe and along the lower edge of each sleeve. Now, slip the garment on your RS, have that bot strut its stuff (Fig. A-4).

Ouch, Something's Wrong

Like any other type of construction project, good sewing comes easily from clean and accurate pattern piece cutting. Likewise, laying the pattern pieces out correctly on the fabric (while observing grain direction and fabric pattern matching) is essential for getting the type of clothes that will make the, err, robot.

Rosebud..., One More Thing

Are you on a roll? Go ahead, make an entire wardrobe of Robosapien clothing. Don't stop with just a robe. A pair of casual pants, a couple of shirts, a fitted jacket—and you'll have one of the best-dressed Robosapiens. Just remember to look for patterns that have been designed for 18-inch dolls. Also, try to avoid synthetic fabrics, which can cause nasty static electricity. Nothing destroys that robotic fashion statement like dragging a couple of dust bunnies in tow behind Robosapien's shuffling feet.

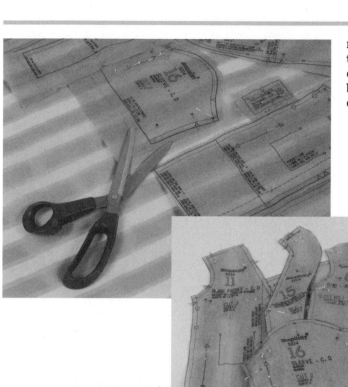

FIGURE A-2 **Lay out the pattern pieces on the fabric, and begin cutting them out.**

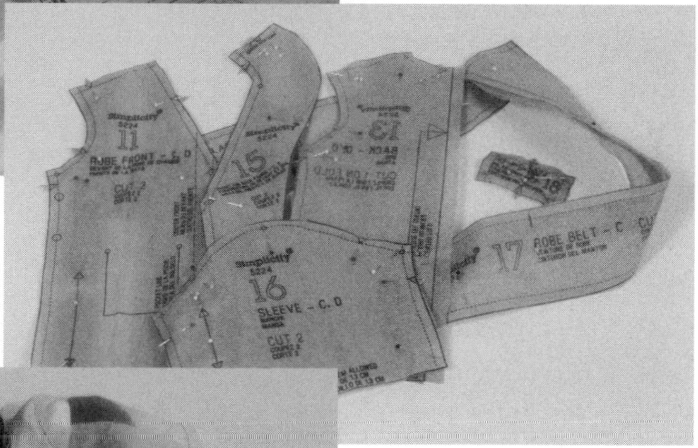

FIGURE A-3 **All of the pieces that make up the Robosapien robe.**

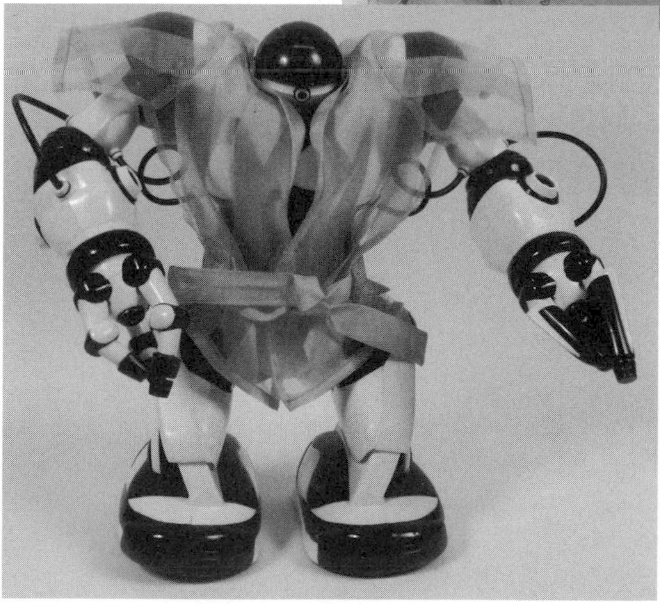

FIGURE A-4 **Get out of here! Those are some pretty sharp threads. The side seams were not sewn together, ensuring the best fit.**

FemBot Is a Male Noun, or Why Girls Hate Robots

As the father of three daughters and devoted husband of an architect, I have a rather unique perspective on hot-button issues like glass ceilings, gender bias, and discrimination. I've suffered and commiserated for years over questionable hiring and firing practices, unequal promotions, and downright laughable management edicts. But if you think that I'm just going to skew off into some tirade about how men are boorish Neanderthals and girls are great, forget it. I just want to know why girls hate robots (Fig. B-1).

Once you accept my motives and appreciate my vantage point, it's tough to totally ignore some of the statistical numbers related to girls and women in science and engineering. For example, according to a 1995 study by the U.S. Department of Education, the percentage of B.S. degrees awarded that year in computer science were 28% for women versus 72% for men. Oddly enough, for advanced degrees, the percentages changed to 26% for women earning M.S. degrees, as opposed to 74% for men, and 18% for women earning doctoral degrees versus 82% for men. What does that mean?

This disparity gets even more dramatic in engineering degrees. Women earned 17% of the B.S. degrees in engineering in 1995, 16% of the M.S.

This article by Dave Prochnow originally appeared in the April 2005 issue of *SERVO Magazine*.

degrees, and 12% of the Ph.D. degrees, while men earned 83%, 84%, and 88%, respectively. Finally, in an oddly titled category of "engineering-related technologies" degrees, women earned 9% of the B.S. degrees, 19% of the M.S. degrees, and 11% of the Ph.D. degrees (with men receiving 91%, 81%, and 89% of those degrees).

Those are some serious differences in degree achievement. Where does that type of incongruity begin? Are there educational factors that contribute to this enormous degree gulf? Or, is there some sort of fundamental attitude that makes us "feel" that someone or some gender is better at some task than another one?

Let's explore each of these questions. First, consider this classic robo-thriller. Are the robots in the 2004 blockbuster movie *I, Robot* (2004, 20th Century Fox) male or female? Your response might be dictated by your own gender or your biases from other science fiction movies. As you'll recall in this movie, Chicago Police Department detective Del Spooner (Will Smith) investigates a threatened social civil war pitting malevolent "free-thinking" robots against the human population. One radical departure for this movie from other sci-fi genre flicks that depict robotic violence is that *I, Robot* featured androgynous beings totally lacking in either the overly testosterone excesses of a Terminator or the evocative sexuality of a *Blade Runner* Pris (Daryl Hannah). These robots were effectively sexless in their physical features. This removal of visual prejudice prevented the audience from anticipating the odd behavior of Sonny (Alan Tudyk) and the other USR robots.

In a National Science Foundation funded research project, investigators Mahzarin Banaji and Anthony Greenwald designed an Implicit Association Test (IAT). Conducted from 1995 to 2002, one aspect of this test was to statistically measure "hidden or unconscious attitudes." Based on the association of words to images, the IAT was able, for example, to demonstrate how people generally attribute higher intelligence and character to attractive men and women based solely on their looks. Makes us sound rather shallow, doesn't it? Is a presidential candidate's physical appearance more important than a campaign platform?

International Flavor

On the Internet, more than 1.5 million international participants took the IAT. The results were sobering. In fact, according to the NSF's report, "people show strong preferences for young versus old, or have far more difficulty grouping women's names with words having to do with science (chemistry, biology), than with arts (drama, poetry). This last result is as apparent in responses of women as in those of men."

Does this mean that we are destined to be racist and sexist? In response, investigator Greenwald states that "you've got these things going on automatically. You may not approve of them, but they're there. And you've got to expect that they might influence your behavior unless you're actively trying to avoid acting in a biased manner."

Girl (Em)Power

Knowledge is power. To empower girls to enter technology fields, you need to arm yourself with some information. Here is a small sampling of Web sites that detail numerous programs and activities that can be used for helping *all* children achieve their potential.

The Advertising Council Math Science, and Technology Education (Issue): www.adcouncil.org/issues/Math_and_Science_Education/

American Association of University Women—Tech-Savvy: Educating Girls in the New Computer Age (2000): www.aauw.org/research/girls_education/techsavvy.cfm

AWE: Attracting Women into Engineering: A Mentoring Program for Middle School Girls: users.rowan.edu/~jahan/personal/kjweb/awe-web/awe.htm

Botswana Roadshow—Girls and Women in Science and Technology: www.col.org/10th/best/botswana.html

Education Development Center: Center for Children and Technology—Telementoring Young Women in Engineering and Computing: Providing the Vital Link: www2.edc.org/CCT/projects_summary.asp?numProjectId=771

EducatingJane.com—Teacher's Resources for Educating Girls: www.educatingjane.com/educators.htm

Girl Scouts Original Research Studies: www.girlscouts.org/research/publications/original/default.asp

IEEE Women in Engineering: www.ieee.org/portal/site/mainsite/menuitem.818c0c39e85ef176fb2275875bac26c8/index.jsp?&pName=corp_level1&path=committee/women&file=index.xml&xsl=generic.xsl

National Science Foundation: Discoveries: www.nsf.gov/discoveries/

Society of Women Engineers: www.swe.org/stellent/idcplg?IdcService=SS_GET_PAGE&nodeId=5

Taiwan News: "Humanoid Robot Dishes Out Kung-Fu Moves": www.etaiwannews.com/Business/2003/11/22/1069466565.htm

Wired News: Bots Not a Bra-Burning Issue: www.wired.com/news/women/0,1540,48337,00.html

So based on the IAT results rather than treating girls like artists, we need to educate them to be technologically and scientifically equal with boys. Oddly enough, Africa is actually ahead of America in realizing this need for "actively trying to avoid acting in a biased manner."

If you don't think that Africa is a growing force in science and technology education, think again; or, better yet, think last century. Yes, it was back in 1989 when the Commonwealth Secretariat Education Programme began soliciting African countries to sponsor a program called a Science and Technology Roadshow for girls.

And in 1990, the Ministry of Education for Botswana became the inaugural host for the Roadshow. Focusing on secondary school students, the program promoted careers in science and technology as a viable option for girls. Unlike later American programs, however, the Roadshow also educated parents, teachers, and even employers about Botswana's need for technology manpower and woman power. Remarkably, predating the IAT results, the Roadshow has mandated that participants should "be encouraged to change their attitudes towards women in these fields."

Although slow to act, the United States is now actively attempting to define the factors that contribute to the gender gulf in technology education. While the initial reaction is to throw money at the topic, one very interesting study in 1999 suggests that filling our classrooms with technology isn't the *sole* answer.

Arguably, the highest tech school system in the United States is the Fairfax County Public School District in Fairfax County, Virginia. Even though this district is the 12th-largest public school district in the country, it is a "technology-rich learning environment" with every classroom wired for the Internet. According to this 1999 study of the 1998 enrollment in all technology classes offered by the Fairfax County Public School District, only 6% of the students enrolled in artificial intelligence classes were young women. Likewise, 23% of the students enrolled in computer programming were female. Interestingly enough, this percentage jumps to near parity with desktop publishing and information system classes, at 46% female enrollment for each subject. Finally, in a rather disappointing result the girls' enrollment (55%) actually exceeded the boys' enrollment in only one class—word processing. This course is considered by some educators to be nothing more than a politically correct euphemism for the typing classes of the mid-20th century.

In engineering-related coursework, the numbers of enrolled girls were even more reduced. One percent of the students in electronics were young women, 15% of the technical drawing students were female, and 18% of the enrolled architectural drawing students were girls. So what are girls in Fairfax County taking? The enrollment figures in foreign language, chemistry, economics, history, art, and music showed a higher percentage of girl participation. The highest girl enrollments, however, where girls outnumbered the boys by 30% or more were in child care, nursing, dance, cosmetology, fashion marketing, dental careers, food occupations, and animal science.

So, are girls just dummies? In some accelerated high schools, like those found in the Fairfax County Public School District, students can take an Advanced Placement (AP) test administered by the College Board and earn college credit in computer science. This is a three-hour exam that typically requires a score of 3 or higher to receive college credit (some schools require a score of 4 or 5 for receiving credit). This AP test is further divided into a one semester introductory exam (i.e., the A exam) and a one-year more comprehensive introductory exam (i.e., the AB exam).

In 1999, 83% of those student who applied for the A exam were male, whereas only 9% of the AB exam test takers were female. Even with such a dramatic difference in the gender ratio, the young women scored higher than their curriculum choices would seem to indicate. For example, this same study showed that while 41% of the girls scored a 1 on the A exam, 17% of them scored a 3; 20%, a 4; and 10%, a 5. And a score of 5 is the highest score that you can achieve. That is an impressive achievement for a group that is traditionally ignored in technology coursework. As for the boys, 19% scored a 3; 25%, a 4; and 17%, a 5. The differences in these scores suggests that girls don't really require a "pink" education but rather a stronger and more supportive foundation in technology that they can build upon (Fig. B-2).

So maybe it's really a matter of packaging. In other words, don't make a whole new curriculum just for girls (those are the educators who think "pink"), instead just modify an existing "girl-appreciated" curriculum and infuse it with computers, programming, robots, and technology. You know the cliché... call it fashion and they will enroll.

Now, before you snort and roll your eyes, maybe you should really study some of today's sewing machines. Although names like Bernina (www.bern-

FIGURE B-2
Making electronics, computers, and robots commonplace in a child's life will prepare her for a happy and healthy life.

ina.com), Janome (www.janome.com), and Pfaff (www.pfaff.com) may not be as familiar to you as Apple, Intel, and Microsoft, these are the big players in the multibillion-dollar fashion industry. Likewise, you might think that your new PowerBook is a high-priced powerhouse; conversely, these sewing machines, like the $5,000+ Pfaff 2144, are incredibly powerful, computerized stitching phenoms that rely on sophisticated programming, the installation of system firmware, and the integration of complex digitizing software for *regular* operation. Can you imagine doing that level of technological work just to write a letter with your word processor? Remarkably, women of all ages are able to perform these operations without any acknowledgment of being a programmer, a system engineer, or a hardware technician. They just like to sew and consider themselves to be fashion designers.

So what can you do? Luckily, you're holding the best tool in your hands right now! *The Official Robosapien Hacker's Guide*, for example, is a great place to start integrating technology into your daughter's life. For example, I read each issue of *SERVO Magazine* to my daughters every month. Sure, they don't really appreciate it at the same level as I do, but they like reading, and they love robots.

Making technology a commonplace household resident will give your children the best foundation that they will ever need. Programming the kitchen's microwave, burning the family video to a DVD, or walking Robosapien around the house help to put a face on technology. Also, watch your children and learn their interests, and then you will be better equipped with integrating technology into their daily lives. And you can't start too early with your introduction to technology. My one-year-old daughter, for example, calls Robosapien "Roro" and can deftly program 14-step activities into its internal memory without assistance (Fig. B-3).

FIGURE B-3
Amelia loves her little "Roro." With just a little instruction, she is now able to program lengthy command sequences into her Robosapien.

But do girls really hate robots? Ironically, it's the inventor of the Robosapien, Mark Tilden, who makes the strongest case for girls liking robots. In a 2003 interview with *Taiwan News*, Tilden offered some insightful quotes

Yes, But What Can *I* Do?

Don't feel helpless, don't give up, and don't just sit there. Here is a brief selection of proactive educational pastimes, practices, and promotions that you can try for bringing technology and robots into children's lives.

BE A FOSTER SCIENTIST. Sure, mentoring is a hot buzzword of wealthy bureaucrats, but it is so hollow (e.g., typical mentoring programs offer a scant 1 or 2 hours of assistance per week). Instead approach mentoring like a foster parent and provide full-time, hands-on science education to kids.

TAKE A ROBOT TO SCHOOL PROGRAM. Use some of your "spare" cash to purchase and donate brand new robots to your local school, church, club, and/or museum. For example, Tamiya America (www.tamiyausa.com) used to make extremely inexpensive robot kits that I have regularly given away. Or, consider a Robosapien for your donation. But please shy away from giving away your old, junk robots. Your school might take them, but the kids will hate them. Also, remember to throw in a couple of buckets worth of batteries—dead batteries or no batteries can make robotics very dull.

TURN THE TV OFF. Although science on TV might be an attractive educational tool, it is a two-dimensional, spoon-fed, in-your-face format that can actually distract a student more than inspire her/him.

SPONSOR SOME SCHOOL SMARTS. Help your local school, church, club, and/or museum purchase an additional computer. Don't try to dictate manufacturer, model, or features, just give them one that matches the administration's requirements.

READ, READ, READ. You can't ever read enough, nor can you begin reading too early. And *SERVO Magazine* is a great source for reading about robots and technology.

regarding a proposed FemBot design (Fig. B-4). Specifically, "She [the proposed FemBot design] can also do simple tasks like combing her owner's hair .... [U]nlike boys, girls like to interact with their toys."

FIGURE B-4 **Just remember: Robosapien —hackable; Barbie—not.**

(*Note:* If you're a robot hacker and you *don't* subscribe to *SERVO Magazine*, then you're just a hack. Subscribe at www.servomagazine.com.)

Addendum: In the cover story (*The Math Myth*) from the March 7, 2005, issue of *Time* magazine, several studies are referenced that suggest that the differences between male and female achievement in science could be attributed to an unpredictable cocktail of environmental influences, social interactions, and human brain physiology. Even more interesting are a pair of studies (both studies are from Iceland) that suggest that girls *and* boys do better in mathematics when they are segregated into female-only and male-only classes. This research is far from conclusive and the jury is still out. All of these studies indicate, however, that more research is needed, *and* we shouldn't just dismiss or trivialize any gender gap as a "girl thing" or a "boy thing."

Robosapien—The Next Generation

O riginally announced at the Consumer Electronics Show (CES) in Las Vegas, Nevada, on January 6, 2005, three new robotic companions were created to help keep Robosapien company. About one month later, WowWee made a more formal announcement (along with some corrected pricing information) about Robosapien's newborn brother (sister) and two new friends from its Hong Kong–based office on February 20, 2005. For your reading enjoyment, here is the complete text of that product announcement:

Prompted by the Success of Robosapien™, WowWee™ Ltd. Introduces a New Line of Entertainment Robots

HONG KONG—February 20, 2005—Following in the "footsteps" of the enormously successful, award-winning Robosapien—the first affordable, highly intelligent pre-programmable entertainment robot—WowWee Ltd. unleashes a line of new robotic companions [Fig. C-1].

Prompted by the proven success of Robosapien and its WowWee Robotics division, the company will expand its platform with a line of Robonetics™ entertainment robots that feature a more sophisticated level of technology.

FIGURE C-1
**Meet your
new family
members.
(Photograph ©
WowWee Ltd.)**

Combining dynamic realistic motion with intelligent technology, WowWee brings its stylish and sleek robotic companions to life with extraordinary features and complex functions sure to engage minds everywhere.

The robots are fully controllable and programmable by remote control and fully autonomous in their free-roam mode. And these highly intelligent robotic companions are able to interact with their owners and one another.

ROBOSAPIEN™ V2

For 2005, WowWee introduces Robosapien V2. Robosapien V2 brings the fluid movement and biomechanical [sic] agility of Robosapien to a whole new level [Fig. C-2]. Gaining a whopping 10" in height, Robosapien V2 now has full range of motion and the ability to pick up, drop and throw objects with his finely tuned precision hands. Advanced agility allows him to bend over and twist from side to side, so he can now sit, bend, lie down and stand up. Robosapien V2 will interact with his surroundings, and even responds with a "real

FIGURE C-2 **I, Robosapien V2, come in peace, so ˜ let's party. (Photograph © WowWee Ltd.)**

voice".... Robosapien V2 can talk! Fully equipped with infrared radar vision, Robosapien V2's moving eyes with blue LED lights can detect obstacles, track movements and take objects handed to him.

Sensory features include a vision color system that enables Robosapien V2 to recognize objects and skin tones; he can wave when he sees you and reach out to shake your hand. A stereo sound detection system allows him to respond and react to noises in the environment. Robosapien V2 comes equipped with "laser" tracking; trace a "laser" path on the ground and he'll follow it. And, Robosapien V2 couldn't carry the Robosapien name if he didn't possess a one-of-a-kind attitude-filled personality! Robosapien V2 is so advanced he can even control his new friends—Roboraptor™ and Robopet™.

AGES: 6 and up
RETAIL: $250
AVAILABLE: December 2005

FIGURE C-3
Watch out Kitty, Roboraptor wants to play. (Photograph © WowWee Ltd.)

ROBORAPTOR™

Dinosaurs may be extinct, but Roboraptor brings them into the age of technology. At 32 inches long, Roboraptor is a realistic robotic beast who is easily controlled and mastered with a remote control [Fig. C-3]. Roboraptor's fluid biomechanical [sic] motion allows for three bipedal movement gaits—predatory, walking and running. Multi-sensor environmental awareness allows Roboraptor to hear, see and feel people and the environment around him—he has multiple touch sensors in the head and tail, while sonic sensors detect sound and direction. Look out—Roboraptor has a strong jaw with powerful snapping action. With three distinct moods: hunter, cautious and playful—Roboraptor will keep you on your toes. Go near his face when he's hunting and he'll behave aggressively; touch his face when he's playful and he'll nuzzle your hand.

AGES: 6 and up
RETAIL: $120
AVAILABLE: August 2005

FIGURE C-4 **Come on Robopet, attack Aibo. (Photograph © WowWee Ltd.)**

ROBOPET™

Robosapien V2 has a new best friend—Robopet! This biomechanical [sic] Robopet is extremely expressive with lifelike animations and digital animal sounds [Fig. C-4]. Robopet is easily programmed with a remote to perform sequences of movements and tricks. Robopet walks, crawls, sits down and stands up, runs and jumps, and can perform an array of pet tricks including lying down, rolling over, begging and howling! Robopet is trainable and responds to positive and negative reinforcement. Multi-sensors allow Robopet to be aware of his surroundings. He is able to avoid obstacles, respond to sounds—even function as a guard dog! When it is time to take Robopet for a walk, use the "laser" light leash on the controller to mark a path for Robopet to follow. Robopet requires seven AAA batteries.

AGES: 6 and up
RETAIL: $90
AVAILABLE: October 2005

WowWee Ltd. is a privately-owned, Hong Kong based company, with offices in North America and a worldwide sales and distribution network. The Company is a recognized leader in the manufacturing of innovative hi-tech consumer electronic and leisure products. Company divisions include WowWee Robotics, WowWee Tech, WowWee RC and WowWee Alive. For more information and to see the latest WowWee products please go to www.wowwee.com.

Things to Know, Places to Go

Here is a collection of every reference, document, and Web site that is mentioned in this book. Live and learn.

Academy Hobby: www.academyhobby.com/

Acroname: www.acroname.com/

All Electronics: www.allelectronics.com/

Article about Mark Tilden by Fred Hapgood that was originally published in *WIRED* in September 1994: www.wired.com/wired/archive/2.09/tilden .html

Article about Mark Tilden by Paul Trachtman that was originally published in the February 2000 issue of *Smithsonian*: www.smithsonianmag.si.edu/ smithsonian/issues00/feb00/robots.html

Article about Robosapien by Marie Feliciano on 22 November 2003 for *Taiwan News*: www.etaiwannews.com/Business/2003/11/22/1069466565.htm

Article about Robosapien dated 12 November 2004 in *The Taipei Times*: www.taipeitimes.com/News/worldbiz/archives/2004/11/12/2003210765

Asimov, Issac, *I, Robot*, Doubleday, New York, 1950

Asimov, Issac, *The Naked Sun*, Doubleday, New York, 1965

Asimov, Issac, *Robot Dreams*, Doubleday, New York, 1986

Asimov, Issac, *Robots and Empire*, Doubleday, New York, 1985

Asimov, Issac, *The Robots of Dawn*, Doubleday, New York, 1983

BASIC Stamp: www.parallax.com/html_pages/products/basicstamps/basic_
stamps.asp

BASICX-24: www.basicx.com/

BG Micro: www.bgmicro.com/

BEAM Online: www.beam-online.com/

"Biomorphic Robots as a Persistent Means for Removing Explosive Mines," from
"Symposium on Autonomous Vehicles in Mine Countermeasures Proceed-
ings," U.S. Naval Postgraduate School, LAUR-95-8411, Spring 1995;
www.fas.org/sgp/othergov/doe/lanl/index2b.html

Bluetooth: www.bluetooth.org/

Botster: www.roboticsconnection.com/r1_robotic_kits_and_platforms.html

BrainStem: www.acroname.com/brainstem/brainstem.html

BRIO: www.brioplay.com/main.asp

C Crane: www.ccrane.com/

ChipCorder: www.winbond-usa.com/products/isd_products/chipcorder/

"Controller for a Four-Legged Walking Machine," by Still and Tilden; Institute
of Neuroinformatics, Zurich, Switzerland, circa 1997; www.solarbotics.net/
library/pdflib/default.htm

DigiKey: www.digikey.com/

EmbeddedBlue: www.a7eng.com/products/embeddedblue/embeddedblue.htm

Erector: www.brioplay.com/Erector_Prod_Template.htm

Eurobot: www.eurobot.org/eng/

Fischer-Technik: www.fischertechnik.com/html/computing-robot-kits.html

Freescale Semiconductor: www.freescale.com/

Gleason Research: www.gleasonresearch.com/

Girder: www.promixis.com/

Handy Cricket: www.handyboard.com/cricket/

iBotz: www.ibotz.com/

"Insectile and Vermiform Exploratory Robots," Thakoor, Kennedy & Thakoor,
NASA Tech Brief, Vol. 23, No. 11, JPL New Technology Report NF'0-2038,
Jet Propulsion Laboratory, Pasadena, California, November 1999; www.
nasatech.com/TSP2/incatyr2.php?cat=Machinery%20-%20Automation
&vol=1999

Interview of Mark Tilden at Bradbury Science Museum in Los Alamos, New Mexico, in March 2000: www.exhibitresearch.com/tilden/

Interview of Mark Tilden in *The New York Times* on 28 November 2004:
www.nytimes.com/2004/11/ 28/magazine/28ROBO.html

iRobot: www.irobot.com/

JoinMax Digital: www.mciirobot.com/ and www.robotplayer.com/

Magnevation: www.magnevation.com/

MCII: www.mciirobot.com/

Me Company: www.mecompany.com/

MEMSIC: www.memsic.com/memsic/

Menzel, Peter, and Faith D'Aluisio, *Robo sapiens: Evolution of a New Species*,
The MIT Press, Cambridge, Mass., 2000; www.mitpress.mit.edu

MGA Entertainment: www.mgae.com/

National Institute of Standards and Technology (NIST): www.nist.gov/

NetMedia: www.netmedia.com/

New Micros: www.newmicros.com/

The Official Robosapien Hacker's Guide website: www.pco2go.com

OWI Robot: www.robotikitsdirect.com/index.html

Parallax: www.parallax.com/

PicoBotz: www.ibotz.com/html/AboutPicoBotz.html

Promixis: www.promixis.com/

Quantum Research Group: www.qprox.com/

RoboFest UK: www.robofesta-europe.org/britain/

Roborior: www.roborior.com/

Robosapien: www.robosapienonline.com/

Robotics Connection: www.roboticsconnection.com/index.html

Robotics Institute: www.ri.cmu.edu/

Roomba: www.irobot.com/consumer/product_detail.cfm?prodid=18

Sanyo: www.sanyo.co.jp/

SpeakJet:www.speakjet.com/

Solarbotics: www.solarbotics.net/

"Theoretical Foundations for Nervous Nets and the Design of Living Machines,"
by Hasslacher and Tilden; Los Alamos National Laboratory Press, Los
Alamos, N.Mex., November 1995; www.solarbotics.net/library/pdflib/
default.htm

Time and Frequency Division (847) of NIST: www.boulder.nist.gov/timefreq/

TiniPod: www.newmicros.com/cgi-bin/store/order.cgi?form=prod_detail&part=
TiniPod

tmsuk: www.tmsuk.co.jp/

WAO Kranius: www.robotikitsdirect.com/products/owi9762.html

Winbond: www.winbond-usa.com/

"Who Says a Woman Can't Be Einstein?" by Ripley; *Time*, March 2005; www.time
.com/time/archive/preview/0,10987,1032332,00.html

WowWee Ltd.: www.wowwee.com/

ZigBee: www.freescale.com/webapp/sps/site/overview.jsp?nodeId=02XPgQhH-
PRjdyB

INDEX